TOPICS IN APPLIED GEOGRAPHY

SOIL EROSION

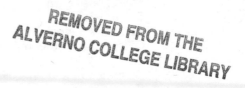

TOPICS IN APPLIED GEOGRAPHY
edited by Donald Davidson and John Dawson

titles published and in preparation:
Slum housing and residential renewal
Soil erosion
Human adjustment to the flood hazard
Office location and public policy
Vegetation productivity
Government and agriculture
Soils and rural land use planning
Physical geography and economic development
in the circumpolar North

R. P. C. Morgan
National College of Agricultural Engineering
Silsoe, Bedfordshire

SOIL EROSION

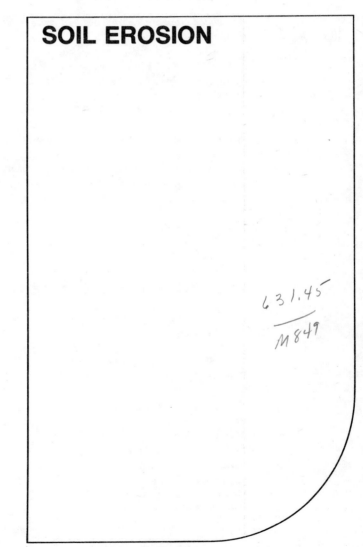

Longman
London
and New York

Longman Group Limited London

*Associated companies, branches and representatives
throughout the world*

*Published in the United States of America
by Longman Inc., New York*

ISBN 0 582 48692 0

First published 1979

Library of Congress Cataloging in Publication Data

Morgan, R P C
  Soil erosion.

  (Topics in applied geography)
  Bibliography: p.
  Includes index.
  1. Soil erosion. 2. Soil conservation. 3. Soil
erosion—Malaysia—Malaya. 4. Soil conservation—
Malaysia—Malaya. I. Title. II. Series.
S623.M68   631.4'5   77-25911
ISBN 0-582-48692-0

Printed in Great Britain by
Richard Clay (The Chaucer Press) Ltd,
Bungay, Suffolk

CONTENTS

Preface   vi

Acknowledgements   viii

Chapter 1   **Distribution of soil erosion**   **1**

Chapter 2   **Processes and mechanics of erosion**   **5**

Chapter 3   **Factors influencing erosion**   **15**

Chapter 4   **Erosion hazard assessment**   **26**

Chapter 5   **Modelling soil erosion**   **41**

Chapter 6   **Strategies for erosion control**   **57**

Chapter 7   **Reconnaissance erosion survey in Peninsular Malaysia**   **70**

Chapter 8   **Semi-detailed erosion survey in central Pahang, Peninsular Malaysia**   **80**

Chapter 9   **Detailed erosion survey and conservation strategy**   **93**

Chapter 10   **Conclusion**   **103**

References   104

Index   111

PREFACE

Soil erosion is a hazard traditionally associated with agriculture in tropical and semi-arid areas. In recent years, however, its importance has become apparent in areas devoted to forestry, transport and recreation. Moreover, erosion is increasingly being recognized as a hazard in temperate countries such as Britain, Belgium and Germany. Conservation measures depend upon a thorough understanding of the mechanics of the erosion processes. This book is the first text to concentrate on soil erosion, a subject which is usually covered only briefly in books on soil conservation.

The mechanics of erosion are reviewed in the first part of the book with emphasis on the extent of and deficiencies in current knowledge. Techniques of classifying land with respect to erosion risk are also examined and a simple working method is presented. A discussion of various approaches to modelling soil erosion focuses on the value of models for predicting rates of soil loss and planning conservation work. Strategies for erosion control are related to the changes that man can make to the soil, plant cover and slope of the land and the effect that these have on the mechanics of erosion. The themes developed in the first part of the book are integrated in a case study of erosion risk evaluation and conservation planning in Peninsular Malaysia. Here the application of erosion mapping, land classification, erosion modelling and conservation systems is described in practice. Attention is given to working with scarce data using remote sensing and detailed field survey.

The book is intended for undergraduate and postgraduate students studying soil erosion and conservation as part of their courses in geography, environmental science, agriculture, agricultural engineering, hydrology, soil science, ecology and civil engineering. In addition it provides an introduction to the subject for those concerned with resources survey and development, agricultural, recreational and rural planning, and for those working on soil erosion and conservation at research or experimental stations.

My thanks are due to the Director, Malaysian Meteorological Service, for supplying many of the data used in the analysis of erosivity for Peninsular Malaysia; Messrs B. B. Lim and C. L. Woo for extracting data from the records of the Meteorological Station of the Department of Geography, University of Malaya; Mr H. H. Soh for assistance with much of the work carried out on the University of Malaya campus; Ing. Zoot. MS. L. Carlos Fierro and Ing. Zoot. J. Jabalera Ramos for help in the field in Mexico; and various students of the Department of Geography, University of Malaya and the National College of Agricultural Engineering, Silsoe who have assisted with both field and laboratory work. The data on soil erosion in Bedfordshire are derived from a research project supported by the Natural Environment Research Council. I am also grateful to my wife, Gillian, for help in the field, reading and improving the manuscript and providing encouragement at all times.

R. P. C. Morgan
Silsoe
September 1977

ACKNOWLEDGEMENTS

We are grateful to the following for permission to reproduce copyright material:

American Society of Agricultural Engineers for Fig. 1 – L. D. Meyer and W. H. Wischmeier (1969) *Transactions* Vol. 12 (6); Edward Arnold (Publishers) Ltd for table from *Drainage Basin Form and Process* by K. Gregory and D. E. Walling (1973); Council of Europe for tables from *Soil Conversation* by M. Fournier – English Edn (1972) Generalstebens Litgrafiska Anstalt for diagram based on Fig. 1 – M. G. Wolman (1976) *Geografiska Annaler* 49 (Series A); Dept. of Geography, Univ. of Singapore for Fig. 8.1 – R. J. Eyles *Journal of Tropical Geography* 25; The Institute of British Geographers for diagram based on Fig. 8 – M. A. Stocking and H. A. Elwell (1967) *Transactions* Vol. 1 (2); McGraw-Hill Book Company for diagram based on Fig. 4.4 from *Streams: Their Dynamics and Morphology* by M. E. Morisawa (1968); Geological Society of Malaysia for table – R. J. Eyles (1970) *Bulletin of Geological Society of Malaysia* Vol. 3; Presses Universitaires de France for Fig. from *Climat et érosion* by F. Fournier (1960); Prentice-Hall Inc for adapted table from *Soil Conservation* by J. H. Stallings (1957). Adapted by permission of publishers; Rhodesia Ministry of Agriculture for Fig. 1 and Table 1 – M. A. Stocking and H. A. Elwell (1973) *Rhodesia Agricultural Journal* 70 (4); T. C. Sheng for table – T. C. Sheng (1972a) *Journal of the Scientific Research Council* Vol. 3 No. 2; Universiti Malaya for data from *Universiti Malaya Metrological Station* 1963–1969 and table – W. P. Panton (1969) from *Natural Resouces in Malaysia and Singapore* ed. B. C. Stone (ex Proc. Symp. Sci. Techn. Res. Malaysia 1967); John Wiley & Sons, Inc for table – G. O. Schwab, R. K. Frevert, T. W. Edminster, K. K. Barnes (1966) from *Soil and Water Conservation Engineering*; Whilst every effort has been made to trace the owners of copyright material in a few cases this has not been possible and we apologise for any rights which we may have unwittingly infringed.

CHAPTER 1
DISTRIBUTION
OF SOIL EROSION

The rapid erosion of soil by wind and water has been a problem since man began cultivating the land. Although it is a less emotive topic today than in the period immediately following the 'Dust Bowl' in the United States in the 1930s, its importance has not diminished. Soil erosion remains a problem in the United States, in many tropical and semi-arid areas and is increasingly recognized as a hazard in temperate countries including Great Britain, Belgium and Germany.

The prevention of soil erosion, which means reducing the rate of soil loss to approximately that which would occur under natural conditions, relies on selecting appropriate strategies for soil conservation and this, in turn, requires a thorough understanding of the processes of erosion. The factors which influence the rate of erosion are rainfall, runoff, wind, soil, slope, plant cover and the presence or absence of conservation measures. In Fig. 1.1 these and other related factors are grouped under three headings: energy, resistance and protection. The energy group includes the potential ability of rainfall, runoff and wind to cause erosion. This ability is termed erosivity. Also included are those factors which directly affect the power of the erosive agents such as the reduction in the length of runoff or wind blow through the construction of terraces and wind breaks respectively. Fundamental to the resistance group is the erodibility of the soil which depends upon its mechanical and chemical properties. Factors which encourage the infiltration of water into the soil and thereby reduce runoff decrease erodibility whilst any activity that pulverizes the soil increases it. Thus cultivation may decrease the erodibility of clay soils but increase that of sandy soils. The protection group focuses on factors relating to the plant cover. By intercepting rainfall and reducing the velocity of runoff and wind, a plant cover protects the soil from erosion. Different plant covers afford different degrees of protection so that, by determining the land use, man, to a considerable degree, can control the rate of erosion.

The rate of soil loss is normally expressed in units of mass or volume per unit area per unit of time. In a review of erosion under natural conditions, Young (1969) quotes rates of the order of $0.004\ 5\ \mathrm{kg\ m^{-2}\ y^{-1}}$ for areas of moderate relief and $0.045\ \mathrm{kg\ m^{-2}\ y^{-1}}$ for steep relief. For comparison, rates from agricultural land, in the range of 4.5 to $45.0\ \mathrm{kg\ m^{-2}\ y^{-1}}$, are classed as accelerated erosion.

Theoretically, whether or not a rate of soil loss is severe is judged relative to the rate of soil formation. If soil properties such as nutrient status, texture and thickness remain unchanged through time it is assumed that the rate of erosion balances the rate of soil formation. Even relatively small changes in the plant cover can result in considerable increases in erosion. Comparing two otherwise identical drainage basins in the highlands of Peninsular Malaysia, Douglas (1967a) shows that in the Telom basin, with 94

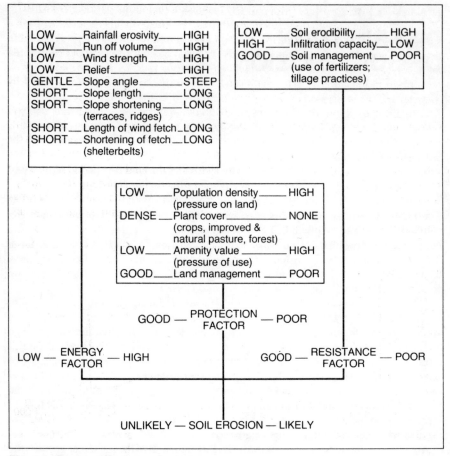

Fig. 1.1 Factors affecting soil erosion.

per cent of its area under rain forest, the erosion rate is 0.21 kg m^{-2} y^{-1}, whilst in the Bertam basin, where only 64 per cent is under forest, the rate is 1.03 kg m^{-2} y^{-1}.

The effects of erosion are felt not only in the areas where top soil is removed by water and wind, the subsoil and bedrock exposed and the land entrenched by gullies, but also in the areas downvalley or downwind where the ground is covered with sand and silt deposits, ditches and canals are clogged with sediment and reservoirs silt up. Deterioration in the quality of cropping and grazing land as a result of erosion brings about reduced productivity and increased expenditure on fertilizers to maintain fertility. In extreme cases yields become so poor that land has to be taken out of cultivation. Siltation of reservoirs and rivers reduces their capacity, creating a flood hazard, and the sediment is a major pollutant, lowering water quality.

The severity of erosion varies in time and space. Most geomorphologists accept that most erosion takes place during events of moderate frequency and magnitude simply because extreme or catastrophic events are too infrequent to contribute appreciably to the quantity of soil eroded over a long period of time. This view is supported by the experimental studies of Roose (1967) in Senegal which show that between 1959 and 1963, 68 per cent of the soil loss took place in rain storms of 15 to 60 mm. Studies of erosion in mid-Bedfordshire (Morgan, 1977) indicate that in the period 1973 to 1975, 99 per cent of the erosion occurred in ten storms, the greatest soil loss resulting from a storm of only 34.9 mm. In contrast, Hudson (1971) emphasizes the role of the more dramatic event. Quoting from research in Rhodesia, he states that 50 per cent of the annual soil loss occurs in only two storms and that, in one year, 75 per cent of the erosion took place in ten minutes. The significance of extreme events is also illustrated by Thornes (1976) in his studies of erosion in southern Spain.

Superimposed on the frequency–magnitude patterns of erosion are changes brought about by man's alterations of the plant cover. A typical sequence of events is described by Wolman (1967) for Maryland where soil erosion rates increased with the conversion of woodland to cropland after AD 1700 (Fig. 1.2). They declined as the urban fringe extended across the area in the 1950s and the land reverted to scrub whilst the farmers sold out to speculators, before accelerating rapidly, reaching 700 kg m^{-2} y^{-1} whilst the area was laid bare during housing construction. With the completion of urban development, runoff from concrete surfaces is concentrated into gutters and sewers, and soil loss falls to below 0.4 kg m^{-2} y^{-1}.

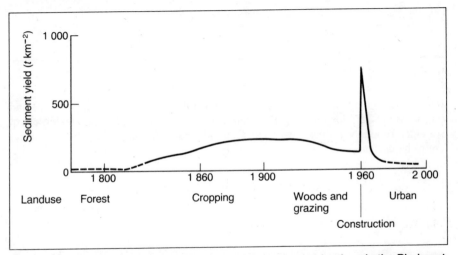

Fig. 1.2 Relationship between sediment yield and changing landuse in the Piedmont region of Maryland (after Wolman, 1967).

Investigations of the relationship between soil loss and climate (Langbein and Schumm, 1958) show that erosion reaches a maximum in areas with a mean annual precipitation of 300 mm. At precipitation totals below this value, erosion increases as precipitation increases. But, as precipitation increases so does the vegetation cover, resulting in better protection of the soil surface. At precipitation totals above 300 mm, the protection effect counteracts the erosive effect of greater rainfall, so that soil loss decreases as precipitation increases. If the vegetation cover is destroyed, however, erosion increases rapidly with precipitation and reaches a maximum in humid tropical areas (Douglas, 1967b).

The relative importance of the factors controlling spatial variations in erosion is dependent upon scale. Using evidence from studies of sediment concentrations in rivers, measurements of soil loss from hillsides, and investigations into the relationships between scale and erosion rates and scale and drainage density, it can be shown that broad regional variations in erosion are attributable to climate, local variations are a

Table 1.1 Factors influencing soil loss at different scales

SCALE OF ANALYSIS			EVIDENCE
Macro	Meso	Micro	
climate	lithology relief		Sediment yield of rivers
climate	lithology relief	micro-climate lithology (soil)	Drainage density
climate	altitude relief		Studies of erosion rates
climate		plant cover micro-climate	Studies of soil loss from hillslopes

After Morgan and Keech (1976).

result of relief, and variations at field level reflect differences in crop cover, slope and the use of conservation measures (Morgan and Keech, 1976; Table 1.1). This pattern of scale-determined variables must be understood if suitable soil conservation strategies are to be designed.

CHAPTER 2
PROCESSES AND
MECHANICS OF EROSION

Soil erosion is a two-phase process consisting of the detachment of individual particles from the soil mass and their transport by erosive agents such as running water and wind. When sufficient energy is no longer available to transport the particles a third phase, deposition, occurs.

Rainsplash is the most important detaching agent. As a result of raindrops striking a bare soil surface, soil particles may be thrown through the air over distances of several centimetres. Continuous exposure to intense rainstorms considerably weakens the soil. The soil is also broken up by weathering processes, both mechanical, by alternate wetting and drying, freezing and thawing and frost action, and biochemical. Soil is disturbed by tillage operations and by the trampling of people and livestock. Running water and wind are further contributors to the detachment of soil particles. All these processes loosen the soil so that it is easily removed by the agents of transport.

The transporting agents comprise those which act areally and contribute to the removal of a relatively uniform thickness of soil and those which concentrate their action in channels. The first group consists of rainsplash, surface runoff in the form of shallow flows of infinite width, sometimes termed sheet flow but more correctly called overland flow, and wind. The second group covers water flow in small channels, known as rills, which can be obliterated by weathering and ploughing, or in the larger, more permanent features of gullies and rivers. To these agents which act externally, picking up material from and carrying it over the ground surface, should be added transport by mass movements such as soil flows, slides and creep, in which water affects the soil internally, altering its strength.

The severity of erosion depends upon the quantity of material supplied by detachment and the capacity of the eroding agents to transport it. Where the agents have the capacity to transport more material than is supplied by detachment, the erosion is described as *detachment-limited*. Where more material is supplied than can be transported, the erosion is *transport-limited*. The recognition of which factor, detachment or transport, is limiting is important because the success or failure of conservation work relies upon applying remedies to the correct one.

The energy available for erosion takes two forms: potential and kinetic. Potential energy (PE) results from the difference in height of one body with respect to another. It is the product of mass (m), height difference (h) and acceleration due to gravity (g), so that

$$PE = mhg \tag{2.1}$$

which, in units of kg, m and m s^{-2} respectively, yields a value in joules. The potential energy for erosion is converted into kinetic energy (KE), the energy of motion. This is related to the mass and the velocity (v) of the eroding agent in the expression

$$KE = \tfrac{1}{2}mv^2 \qquad\qquad (2.2)$$

which, in units of kg and (m s^{-1})^2, also gives a value in joules. Most of this energy is dissipated in friction with the surface over which the agent moves so that only 3 to 4 per cent of the energy of running water and 0.2 per cent of that of falling raindrops is expended in erosion (Rubey, 1952; Pearce, 1976). An indication of the relative efficiencies of the processes of water erosion can be obtained by applying these figures to calculations of kinetic energy, using equation (2.2), based on typical velocities (Table 2.1). The concentration of running water in rills affords the most powerful erosive agent but raindrops are potentially more erosive than overland flow. Most of the raindrop energy is used in detachment, however, so that the amount available for transport is less than that from overland flow. This is illustrated by measurements of soil loss in the field in mid-Bedfordshire. Over a 900-day period, on an 11° slope on a sandy soil, transport across a centimetre width of slope amounted to 19 000 g of sediment by rills, 400 g by overland flow and only 20 g by rainsplash (Morgan, 1977).

Table 2.1   Efficiency of forms of water erosion

FORM	MASS*	TYPICAL VELOCITY (m s^{-1})	KINETIC ENERGY†	ENERGY FOR EROSION§	OBSERVED SEDIMENT TRANSPORT‖ (g cm^{-1})
Raindrops	R	9	$40.5R$	$0.081R$	20
Overland flow	$0.5R$	0.01	$2.5\times10^{-5}R$	$7.5\times10^{-7}R$	400
Rill flow	$0.5R$	10	$25R$	$0.75R$	19 000

  * Assumes rainfall of mass R of which 50 per cent contributes to runoff.
  † Based on $\tfrac{1}{2}mv^2$.
  § Assumes that 0.2 per cent of the kinetic energy of raindrops and 3 per cent of the kinetic energy of runoff is utilized in erosion.
  ‖ Totals observed in mid-Bedfordshire on an 11° slope, on sandy soil, over 900 days. Most of the energy of raindrops contributes to detachment rather than transport.

2.1. RAINSPLASH EROSION

The action of raindrops on soil particles is most easily understood by considering the momentum of a single raindrop falling on a sloping surface. The downslope component of this momentum is transferred in full to the soil surface but only a small proportion of the component normal to the surface is transferred, the remainder being reflected. The transference of momentum to the soil particles has two effects. First, it provides a consolidating force, compacting the soil, and second, it imparts a velocity to some of the soil particles, launching them into the air. On landing, they transfer their own downslope momentum to other particles and the jumping process is repeated. Thus, raindrops are agents of both consolidation and dispersion.

The consolidation effect is best seen in the formation of a surface crust, usually only a few millimetres thick, which results from the clogging of the pores by soil compaction. It has been suggested by Young (1972) that this is associated with the dispersal of fine particles from soil aggregates or clods which are translocated to infil the pores. Laboratory studies by Farmer (1973), however, show that it is the medium and coarse particles that are most easily detached from the soil mass and that clay particles resist detachment. This may be because the raindrop energy has to overcome the adhesive or chemical bonding forces by which the minerals comprising clay particles are linked (Yariv, 1976). One way in which the dispersion of soil particles occurs, in addition to the direct impact of falling raindrops, is by slaking (Bryan, 1969). This is the breakdown of the soil by the compression of air ahead of a wetting front as rainfall starts to infiltrate a dry soil from the surface downwards.

Rain does not always fall on to a dry surface. During a storm it may fall on surface water in the form of puddles or overland flow. Studies by Palmer (1964) show that as the thickness of the surface water layer increases, so does splash erosion. This is believed to be due to the turbulence which impacting raindrops impart to the water. There is, however, a critical water depth beyond which erosion decreases again because all the energy is dissipated in the water and does not affect the soil surface. Laboratory experiments have shown that the critical depth is approximately equal to the diameter of the raindrops.

Because splash erosion acts uniformly over the land surface its effects are seen only where stones or tree roots selectively protect the underlying soil and splash pedestals or soil pillars are formed. Such features frequently indicate the severity of erosion. On sandy soils in Bedfordshire 2 cm high pedestals can form in one year (Morgan, 1977).

2.2. OVERLAND FLOW

Overland flow occurs on hillsides during a rainstorm when surface depression storage and either, in the case of prolonged rain, soil moisture storage or, with intense rain, the infiltration capacity of the soil are exceeded. The flow is rarely in the form of a sheet of water of uniform depth and more commonly is a mass of anastomosing or braided water courses with no pronounced channels. The flow is broken up by large stones and cobbles and by the vegetation cover, often swirling around tufts of grass and small shrubs.

The hydraulic characteristics of the flow are described by its Reynolds number (Re) and its Froude number (F), defined as follows:

$$Re = \frac{vr}{\nu} \tag{2.3}$$

$$F = \frac{v}{\sqrt{gr}} \tag{2.4}$$

where r is the hydraulic radius which, for overland flow, is taken as equal to the flow depth and ν is the kinematic viscosity of water. The Reynolds number is an index of the turbulence of the flow. The greater the turbulence, the greater is the erosive power generated by the flow. Thus, as has been shown experimentally in the laboratory (D'Souza and Morgan, 1976), soil loss varies with the Reynolds number. At numbers less than 500 laminar flow prevails and at values above 2 000 flow is fully turbulent. Intermediate values are indicative of transitional or disturbed flow, often a result of turbulence being imparted to laminar flow by raindrop impact (Emmett, 1970). When the Froude number is less than 1.0 the flow is described as tranquil or subcritical; values greater than 1.0 denote supercritical or rapid flow which is the more erosive. According to the laboratory studies Savat (1977) most overland flow is supercritical and Froude numbers can be as high as 15. Field studies of overland flow in Bedfordshire reveal Reynolds numbers between 1 and 50 and Froude numbers between 0.01 and 0.1.

The important factor in these hydraulic relationships is the flow velocity. Because of an inherent resistance of the soil, velocity must attain a threshold value before erosion commences (Hjulström, 1935). As can be seen in Fig. 2.1, for grains larger than 0.5 mm in diameter, the critical velocity increases with grain size. A larger force is required to move larger particles. For grains smaller than 0.5 mm, the critical velocity increases with decreasing grain size. The finer particles are harder to erode because of the cohesiveness of the clay minerals which comprise them. Once an individual grain is in motion, it is not deposited until velocity falls below the fall velocity threshold. Thus, less force is needed

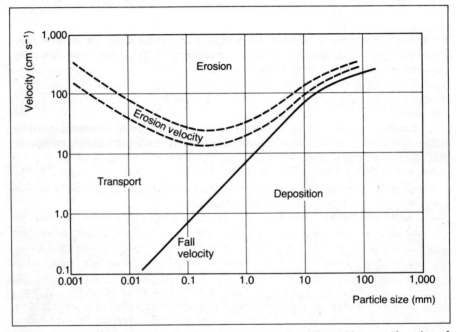

Fig. 2.1 Critical water velocities for erosion, transport and deposition as a function of particle size (after Hjulström, 1935).

to keep a grain in motion than to entrain it. A soil particle of 0.01 mm requires a flow of 60 cm s^{-1} to detach it but it is not deposited until the flow velocity falls below 0.1 cm s^{-1}. In practice, the actual velocities required to erode soil differ from the values shown in Fig. 2.1 because the latter are derived for surfaces of uniform grain size. With a mixed grain size, the finer particles are protected by the coarser ones so that they are not removed until velocity is great enough to pick up the larger grains. Counteracting this effect, however, particularly in shallow flows, is the action of rainsplash which may detach soil particles and throw them into the flow. They are then transported until the flow velocity becomes less than the required fall velocity.

The velocity of flow is dependent upon the flow depth or hydraulic radius, the roughness of the surface and the slope (S). This relationship is commonly expressed by the Manning equation:

$$v = \frac{r^{2/3} S^{1/2}}{n} \tag{2.5}$$

where slope is a real number (m m^{-1}), though tan θ or sin θ, where θ is the gradient angle may be substituted on gentle slopes, and n is the Manning coefficient of roughness. Equation 2.5 assumes that the flow is fully turbulent and moving over a rough surface. Where flow does not satisfy this condition, the exponents in the equation change their values and Savat (1977) recognizes the following variations with Reynolds number:

$$v \propto r^{1.7} \ S^{0.95} \quad \text{for } Re = \quad 250 \tag{2.6}$$
$$v \propto r^{0.95} S^{0.7} \quad \quad Re = \quad 500 \tag{2.7}$$
$$v \propto r^{0.5} \ S^{0.4} \quad \quad Re = 1\,000. \tag{2.8}$$

Using the continuity and Manning's velocity equations Meyer (1965) shows that

$$v \propto S^{1/3} Q_w^{1/3} \tag{2.9}$$

for constant roughness conditions, where Q_w is the water discharge or flow rate. Assuming that the detachment (Df) and transporting (Tf) capacities of flow vary respectively with the square and fifth power of velocity:

$$Df \propto S^{2/3} Q_w^{2/3} \tag{2.10}$$
$$Tf \propto S^{5/3} Q_w^{5/3} \tag{2.11}$$

(Meyer and Wischmeier, 1969). The last equation compares closely with that derived by Carson and Kirkby (1972) from a consideration of the hydraulics of sediment transport:

$$Qs = 0.015\,8\ Q_w^{1.75} k^{-1.11} (\sin \theta \cos \theta)^{1.625} \tag{2.12}$$

where Qs is the sediment discharge per unit width and k is the particle size of the surface material at which 84 per cent of the grains are finer.

It can be seen from the above that the amount of soil loss resulting from erosion by overland flow varies with the velocity and the turbulence of flow. Equally important, however, is the spatial extent of the flow. Horton (1945) describes overland flow as covering two-thirds or more of the hillsides in a drainage basin during the peak period of a storm. The flow results from the intensity of the rainfall being greater than the infiltration capacity of the soil and is distributed over the hillslopes in the following pattern. At the top of a slope is a zone without flow which forms a belt of no erosion. At a critical distance from the crest sufficient water has accumulated on the surface for flow to begin. Moving further downslope the depth of flow increases with distance from the crest until, at a further critical distance, the flow becomes channelled and breaks up into rills. That overland flow occurs in such a widespread fashion has since been questioned, particularly in well-vegetated areas where such flow occurs infrequently and covers only that 10 to 30 per cent of the area of a drainage basin closest to the stream sources (Kirkby, 1969a). Under these conditions its occurrence is more closely related to saturation of the soil and the fact that soil moisture storage capacity is exceeded rather than infiltration capacity. Although, as illustrated by the detailed studies of Dunne and Black (1970) in a small forested catchment in Vermont, the saturated area expands and contracts, being sensitive to heavy rain and snow melt, rarely can erosion by overland flow affect more than a small part of the hillslopes.

Since most of the observations testifying to the power of overland flow relate to semi-arid conditions or to areas with sparse plant cover, it would appear that vegetation is the critical factor. Some form of continuum exists, ranging from well-vegetated areas where overland flow occurs rarely and is mainly of the saturation type, to bare soil where it frequently occurs and is of the Hortonian type. Removal of the plant cover can therefore enhance erosion by overland flow. The change from one type of overland flow to another results from more rain reaching the ground surface, less being intercepted by the vegetation, and decreased infiltration as rainbeat causes a surface crust to develop.

2.3. SUBSURFACE FLOW

Much attention in recent hydrological research has been paid to the lateral movement of water downslope through the upper layers of the soil. Where this takes place as concentrated flow in tunnels or subsurface pipes its erosive effects through tunnel collapse and gully formation are well known. Less is known about the eroding ability of water moving through the pore spaces in the soil, although it has been suggested that fine particles may be washed out by this process (Swan, 1970; Morgan, 1973). Field experiments by Burykin (1957) showed that fines could be moved mechanically through the voids in the soil but under the somewhat extreme conditions in which simulated rain was applied to 20° to 25° slopes at a rate of 2.5 mm min^{-1} for 1 h.

More realistic studies by Roose (1970) in Senegal reveal that soil water flow contributes only about 1 per cent of the total material eroded from a hillside and that this is mainly in the form of colloids and minerals in ionic solution. Subsurface flow is more important, however, than this figure suggests because the concentrations of base minerals in the water are twice those found in surface flow. Essential plant nutrients, particularly those added in fertilizers, can be removed by this process, thereby impoverishing the soil and reducing its resistance to erosion.

2.4. RILL EROSION

Most soil erosion studies do not distinguish the effects of rill erosion from those of overland flow. Both processes generally affect the same part of a hillslope and their hydraulic characteristics are similar (Mosley, 1974). Rills are also ephemeral features. Those from one storm are often obliterated before the next storm of sufficient intensity to cause rilling, when the channels may form an entirely fresh network, unrelated to the positions of previous rills. Most rill systems are discontinuous, that is they have no connexion with the main river system. Only occasionally does a master rill develop a permanent course with an outlet to the river.

As outlined earlier, it is widely accepted that rills are initiated at a critical distance downslope where overland flow becomes channelled. Observations in mid-Bedforshire suggest that this need not always be the case. Here rills develop from a sudden burst of water on to the surface near the bottom of the slope where a small cut is formed which rapidly extends headwards upslope as a channel (Morgan, 1977). Such bursts may well be associated with saturation overland flow rather than flow of the Hortonian type.

As expected from the greater erosive power of concentrated flow, rill erosion may account for the bulk of the sediment removed from a hillside. Studies in the United States on slope plots 4.5 m long show that over 80 per cent of the sediment is transported in rills (Mutchler and Young, 1975). Part of this material is derived from the inter-rill areas and is moved into the rills by overland flow and rainsplash. Foster and Meyer (1975) estimate that up to 87 per cent of the sediment transported by rills may be derived in this way. In their study of rill erosion near Oudenaarde, Belgium, Gabriels, Pauwels and De Boodt (1977) estimate that half of the material removed from the inter-rill areas by overland flow is washed into the rills. A major factor determining the importance of rill erosion is the spacing of the rills. At Oudenaarde, rills occur at 8 to 9 m intervals across the slope whereas, in mid-Bedfordshire, they occur at only 300 to 350 m intervals and over 90 per cent of the soil loss is attributable solely to overland flow (Morgan, 1977).

2.5. GULLY EROSION

Gullies are relatively permanent steep-sided water courses which experience ephemeral flows during rainstorms. They are almost always associated with accelerated erosion and therefore with landscape instability. At one time it was thought that gullies developed as enlarged rills but studies in the gullies or arroyos of the southwest United States revealed that their initiation is a more complex process. In the first stage small depressions or knicks form on a hillside as a result of localized weakening of the vegetation cover by grazing or by fire. Water concentrates in these depressions and enlarges them until several depressions coalesce and an incipient channel is formed. Erosion is concentrated at the heads of the depressions where near-vertical scarps develop over which supercritical flow occurs. Some soil particles are detached from the scarp itself but most erosion is associated with scouring at the base of the scarp which results in deepening of the channel and undermining of the headwall, leading to collapse and retreat of the scarp upslope (Ologe, 1972). Sediment is also produced further down the gully by streambank erosion. This occurs partly by the scouring action of running water and the sediment it contains and partly by slumping of the banks following saturation during flow. Between flows sediment is made available for erosion by weathering and bank collapse. This sequence of gully formation (Fig. 2.2), described by Leopold, Wolman and Miller (1964) in New Mexico, has also been observed in New Zealand (Blong, 1970) and in the New Forest in England (Tuckfield, 1964).

Not all gullies develop purely by surface erosion, however. Berry and Ruxton (1960), investigating gullies in Hong Kong which formed following clearance of the natural forest cover, found that they could not be attributable to surface flow. Most water was removed from the hillsides by subsurface flow in pipes and when heavy rain provided sufficient flow to flush out the soil in these, the ground surface subsided, exposing the pipe network as gullies. Numerous studies record the formation of gullies by pipe or tunnel collapse. It has been observed in the Sudan (Berry, 1970), on loessial soils in the United States (Buckham and Cockfield, 1950) and in Hungary (Zaborski, 1972), and on sodic soils in the western United States (Heede, 1971). Another way in which gullies are initiated is where linear landslides leave deep, steep-sided scars which may be occupied by running water in subsequent storms. This type of gully development has been described in Italy by Vittorini (1972).

When designing strategies to control gully erosion it is not possible to treat all gullies in the same way. Indeed, the dangers of so doing are inherent in the failure to take account of whether surface or subsurface erosion is the major cause. Differences between gullies become even more marked when networks rather than individual channels are considered. Three types of network can be recognized, described by De Ploey (1974), based on studies in the Kasserine Basin, Tunisia, as axial, digitate and frontal. The types are related to differences in soils and the effects that these have on the processes of gully formation. *Axial gullying,* which consists of individual gullies with single headcuts that retreat upslope by surface erosion, occurs in gravelly deposits. *Digitate gullying,* where retreat occurs in several headcuts extending in the direction of tributary depressions, is characteristic of clay loams. *Frontal gullying* is associated with piping and is found particularly on loamy sands with columnar structure. This latter type generally starts from river banks where pipes have their outlet and collapse ensues.

2.6. MASS MOVEMENTS

Although mass movement has been widely studied by geologists, geomorphologists and

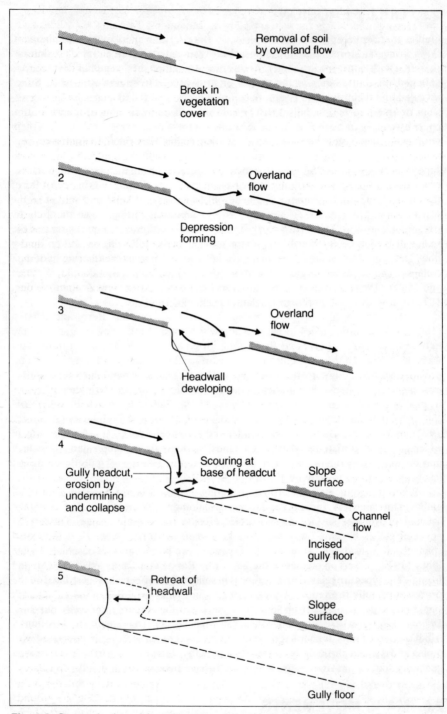

Fig. 2.2 Stages in the surface development of gullies on a hillside.

engineers it is generally neglected in the context of soil erosion. Yet Temple and Rapp (1972) have found that in the western Uluguru Mountains, Tanzania, landslides and mudflows are the dominant erosion processes. They occur in small numbers every wet season and in large numbers about once every ten years. The quantity of sediment moved from the hillsides into rivers by mass movement is far in excess of that contributed by gullies, rills and overland flow. Further, less than 1 per cent of the slide scars are in areas of woodland, 47 per cent being on cultivated plots and another 47 per cent on land lying fallow. The association of erosion with woodland clearance for agriculture is thus very clear.

Mass movements, in the varied forms of creep, slides, rock falls and mudflows, are given detailed treatment in numerous books (Sharpe, 1938; Leopold, Wolman and Miller, 1964; Zaruba and Mencl, 1969) but, for soil erosion studies, rather than stress the separate forms, it is more helpful to consider them as part of a continuum of flow phenomena, ranging from debris slides, in which the ratio of solid to liquid is high, through mudflows to running water which has a low solid–liquid ratio. The close relationship between mass movement and water erosion is illustrated by the studies of the so-called bottle slides in the Uluguru Mountains (Temple and Rapp, 1972; Lundgren and Rapp, 1974) which develop in areas of large subsurface pipes as a result of the flushing out of a muddy viscous mass of debris and subsequent ground collapse. The scars of these slides are similar to the pear-shaped ravines called lavakas, found in the Malagasy Republic, which, in turn, are not unlike gullies (Tricart, 1972).

2.7. WIND EROSION

The entrainment of soil particles by wind is effected by the application of a sufficiently large fluid force and by the bombardment of the soil by grains already in motion. Recognizing these two forces, Bagnold (1937) identifies two threshold velocities required to initiate grain movement. The static or fluid threshold applies to the direct action of the wind and the dynamic or impact threshold allows for the bombarding effect of moving particles. The critical velocities vary with the grain size of the material, being least for particles of 0.10 to 0.15 mm in diameter and increasing with both increasing and decreasing grain size (Chepil, 1945). The resistance of the larger particles results from their size and weight. That of the finer particles is due to their cohesiveness and the protection afforded by surrounding coarser grains (Fig. 2.3).

The transport of soil and sand particles by wind takes place in suspension, surface creep and saltation. Suspension describes the movement of fine particles, usually less than 0.2 mm diameter, high in the air and over long distances. Surface creep is the rolling of coarse grains along the ground surface. Saltation is the process of grain movement in a series of jumps. The movement is explained by considering the theory of the Bernoulli effect. After a grain has been rolling along the ground for a short distance, the velocity of the air at any point near the grain is made up of two components, one due to the wind and the other to the spinning of the grain. On the upper side of the grain both components have the same direction but on the lower side are in opposite directions. As a result of greater velocity on the top surface of the grain, the pressure there is reduced whilst pressure at the lower surface increases. This difference in static pressure produces a lifting force and when this is sufficient to overcome the weight of the grain, the grain rises vertically. The fall trajectory has an angle of 6° to 12°. On striking the ground surface the impact energy of a saltating grain is distributed into a disruptive part which causes disintegration of the soil and a dispersive part which imparts a velocity to the soil

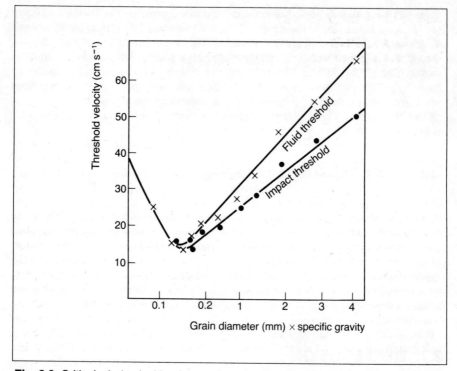

Fig. 2.3 Critical wind velocities for erosion as a function of particle size and specific gravity.

particles and launches them into the air (Smalley, 1970). In a soil blow between 55 and 72 per cent of the moving particles are carried in saltation.

Laboratory studies in wind tunnels show that the detachment and transport capacities of wind vary with the square and the cube of the velocity respectively. Developing the second of these relationships, Chepil (1945) obtains the following equation to describe the sediment discharge ($Qs W$) per unit width ($kg\ m^{-1}\ h^{-1}$) for grains of 0.25 mm diameter:

$$Qs W = 52(V - Vt)^3,\qquad\qquad(2.13)$$

where V is the wind velocity ($cm\ s^{-1}$) measured at a height of 1 m, and Vt is the threshold wind velocity, usually taken as 400 cm s^{-1}, as measured at 1 m height, required to initiate particle movement at the ground surface.

Although an understanding of the mechanics of erosion is vital to the design of a suitable system of erosion control, additional information is necessary for the successful planning of soil conservation works. Account must be taken of the severity of erosion, the frequency at which each process operates and the size of the area affected. Attention to these details requires a study of the factors influencing the soil erosion system.

CHAPTER 3
FACTORS
INFLUENCING EROSION

The factors controlling the working of the soil erosion system are the erosivity of the eroding agent, the erodibility of the soil, the slope of the land and the nature of the plant cover. In order to understand when and how much erosion is likely to occur these factors must be examined in detail and the relevant aspects of them identified more precisely. Much of our understanding stems from empirical studies in which a wide range of data on soil loss and presumed controlling variables is collected and the best relationships are sought using statistical techniques, particularly correlation and regression analyses. Since this approach is adopted by numerous researchers working in many different areas of the world, it is not too surprising that the result is a multiplicity of variables being recognized as important but some disagreement as to which are the most significant. In a review of the relationship between erosion and climate, Douglas (1976) lists seven precipitation variables which have been used to explain spatial variations in erosion and a further seven for precipitation and eight describing antecedent conditions used to study temporal variations in erosion at single sites. In trying to simplify such lists, there is the problem of determining which variables merely express the same relationship and which identify truly separate relationships with soil loss. In this chapter, only those variables which are commonly accepted as important are discussed.

3.1. EROSIVITY

3.1.1. Rainfall

Soil loss is closely related to rainfall partly through the detaching power of raindrops striking the soil surface and partly through the contribution of rain to runoff. This applies particularly to erosion by overland flow and rills for which intensity is generally considered to be the most important rainfall characteristic. The effect of rainfall intensity is illustrated by the data for 183 rain events which caused erosion at Zanesville, Ohio, between 1934 and 1942 which show that average soil loss per rain event increases with the intensity of the storm (Table 3.1; Fournier, 1972).

The role of intensity is not always so obvious, however, as indicated by studies of erosion in mid-Bedfordshire. Taking data for the ten most erosive storms between May 1973 and October 1975 (Table 3.2), it can be seen that whilst intense storms, such as the one on 6 July 1973 of 34.9 mm, in which 17.7 mm fell at intensities greater than 10 mm h^{-1}, produce erosion so do storms of long duration and low intensity, like the one of 19 June 1973 when 39.6 mm of rain fell in over 23 h (Morgan, 1977). A similar picture emerges from measurements of erosion made in 1975 in vineyards near Trier, West

Table 3.1   Relationship between rainfall intensity and soil loss

MAXIMUM 5-MIN INTENSITY (mm h^{-1})	NUMBER OF FALLS OF RAIN	AVERAGE EROSION PER RAINFALL (kg m^{-2})
0– 25.4	40	0.37
25.5– 50.8	61	0.60
50.9– 76.2	40	1.18
76.3–101.6	19	1.14
101.7–127.0	13	3.42
127.1–152.4	4	3.63
152.5–177.8	5	3.87
177.9–254.0	1	4.79

Data for Zanesville, Ohio, 1934–42 (after Fournier, 1972).

Table 3.2   Rainfall and soil loss in Bedfordshire

DATE	RAINFALL (mm)	INTENSE RAIN ($\geqslant$10 mm h^{-1}) (mm)	RAINFALL DURATION (h)	SOIL LOSS (g cm^{-1})
6.7.73	34.9	17.7	8.00	70.70
5.5.73	6.7	—	4.16	61.72
21.5.73	7.1	3.9	2.00	55.98
27.8.73	16.8	4.6	6.50	50.90
6.5.73	2.2	1.0	0.33	40.00
8.8.74	17.0	8.8	13.50	36.78
16.7.74	7.6	6.4	13.00	28.66
13.7.74	16.5	9.0	10.50	24.71
19.6.73	39.6	—	23.83	20.38
27.6.73	18.6	2.3	13.83	18.99

Data for bare soil and 11° slope.

Germany (Richter and Negendank, 1977) where a storm of 15.6 mm with a maximum intensity of 50.4 mm h^{-1} resulted in soil loss from a 26° slope of 8 m length of 141 g m^{-1} whilst a storm of 19.8 mm and a maximum intensity of 44.4 mm h^{-1} caused a greater loss of 242 g m^{-1}. Further, a storm of 39 mm spread over two days with a maximum intensity of 25.8 mm h^{-1} produced a loss of 27 g m^{-1} but one of 30.8 mm and 31.2 mm h^{-1} yielded only 17 g m^{-1}. It appears that erosion is related to two types of rain event, the short-lived intense storm where the infiltration capacity of the soil is exceeded and the prolonged storm of low intensity which saturates the soil. In many instances it is difficult to separate the effects of these two types of event in accounting for soil loss.

The response of the soil in terms of erosion to the receipt of rainfall may be determined by previous meteorological conditions. This can again be demonstrated by data for Zanesville, Ohio (Table 3.3; Fournier, 1972). Over the period 9 to 18 June 1940, the first rain fell on dry ground and, in spite of the quantity of rain, little runoff resulted, most of the water soaking into the soil. In the second storm, however, 66 per cent of the rain ran off and soil loss almost trebled. The control in this case is the

Table 3.3 Influence of antecedent rainfall conditions on soil loss

DATE	RAINFALL (mm)	RUNOFF (% of rainfall)	EROSION (g m^{-2})
9 June	19.3	25	1.5
10 June	13.7	66	4.0
11 June	23.8	69	8.9
15 June	14.0	65	4.2
17–18 June	13.0	50	4.6

Data for Zanesville, Ohio, June 1940, for five successive rainstorms on a plot of 20 m^2 (after Fournier, 1972).

closeness of the soil to saturation which is dependent on how much rain has fallen in the previous few days. The pattern of low soil loss in the first and high loss in the second of a series of storms is reversed, however, where, between erosive storms, weathering and light rainfall loosen the soil surface. Most of the loose material is removed during the first runoff event leaving little for erosion in subsequent events. This sequence is illustrated by studies in the Alkali Creek watershed, Colorado (Heede, 1975) where a sediment discharge peak of 143 kg s^{-1} with a runoff from snowmelt of 2.21 m^3 s^{-1} was observed in one of the ephemeral gullies on 15 April 1964. This event followed one year without runoff. Next day peak runoff increased to 3.0 m^3 s^{-1} but the sediment discharge fell to 107 kg s^{-1}. Although this type of evidence clearly points to the importance of antecedent events in conditioning erosion, other observations deny its significance. No relationship was obtained between soil loss and antecedent precipitation in mid-Bedfordshire (Morgan, 1977) nor between precipitation, soil loss and soil moisture at the onset of rain in the Trier area (Richter and Negendank, 1977).

The question arises of how much rain is required to induce significant erosion. In Chapter 1 it was noted that whilst the importance of extreme events is emphasized by some observers, most workers agree that most erosion occurs in moderate events exemplified by rain storms yielding 30 to 60 mm. Similar opinions are evident in attempts to define a critical rainfall intensity for erosion. Hudson (1971) gives a figure, based on his studies in Rhodesia, of 25 mm h^{-1}, a value which has also been found appropriate in Tanzania (Rapp, Axelsson, Berry and Murray-Rust, 1972) and in Malaysia (Morgan, 1974). It is too high a value for western Europe, however, where it is only rarely exceeded. Arbitrary thresholds of 10 mm h^{-1} and 6 mm h^{-1} have been used in England (Morgan, 1977) and West Germany (Richter and Negendank, 1977) respectively.

Threshold values vary with the erosion process. The figures quoted above are typical for erosion by overland flow and rills, as indicated by numerous observations. Overland flow occurs in the Kuala Lumpur area of Malaysia with rainfall intensities of 60 to 75 mm h^{-1} (Morgan, 1972). Rill erosion has been reported in Cambridgeshire following a storm of 7.4 mm (Evans and Morgan, 1974) and in Belgium following a month in which 213 mm of rain fell but with totals over 20 mm on only two days (Gabriels, Pauwels and De Boodt, 1977). The return period of overland flow at Kuala Lumpur is 60 days and, whilst insufficient data exist to determine the equivalent return period for eastern England, a rainstorm of 37 mm in Bedfordshire has a recurrence interval of 5 years. That significant erosion can result from only moderate events also applies to landslides. Many of those described by Temple and Rapp (1972) in the Mgeta area of the Uluguru Mountains occurred in a storm of 100.7 mm in three hours on 23

February 1970, an event with a return period there of only 4.6 years.

These rainfall totals and return periods are of a different magnitude from those required to initiate gullying. Instances of fresh gullying have been described in the Appalachian Mountains (Hack and Goodlett, 1960), the Alps (Tricart, 1961), the Kuantan area of Malaysia (Nossin, 1964) and in many parts of eastern Europe (Starkel, 1976) and all relate to events with return periods in excess of 10 years. Although gully erosion and erosion by overland flow are generally associated with meteorological events of different frequencies and magnitudes, the distinction between them is occasionally blurred. This occurs in those areas which, by world standards, regularly experience what may be described as extreme events. Starkel (1972) stresses the importance of regular gully erosion in the Assam Uplands where monthly rainfall may total 2 000 to 5 000 mm and in the Darjeeling Hills where over 50 mm of rain falls on an average of twelve days each year and rainfall intensities are often highest at the end of a rain event. A further problem is that the effects of an extreme event may be long lasting and give rise to high soil losses for a number of years. The length of time required for an area to recover from a severe rainstorm, flooding and gullying has not been fully investigated but in a review of somewhat sparse evidence Thornes (1976) quotes figures up to 50 years.

3.1.2. Rainfall erosivity indices

The most suitable expression of the erosivity of rainfall is an index based on the kinetic energy of the rain. Thus the erosivity of a rainstorm is a function of its intensity and duration, and of the mass, diameter and velocity of the raindrops. To compute erosivity requires an analysis of the drop-size distributions of rain. Laws and Parsons (1943), based on studies of rain in the eastern United States, show that drop-size characteristics vary with the intensity of the rain; for example, the median drop diameter (d_{50}) increases with rainfall intensity (Fig. 3.1). Studies of tropical rainfall (Hudson, 1963) indicate that this relationship holds only for rainfall intensities up to 100 mm h^{-1}. At greater intensities median drop size decreases with increasing intensity, presumably because greater turbulence makes larger drop sizes unstable. More recent investigations of rainfall in temperate latitudes (Mason and Andrews, 1960; Carter, Greer, Braud and Floyd, 1974) reveal that the relationship between median drop size and intensity is not

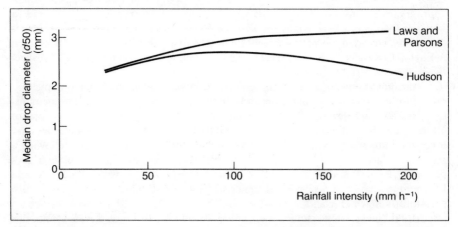

Fig. 3.1 Relationship between median drop diameter and rainfall intensity (after Hudson, 1965).

constant but that both median drop size and drop-size distribution vary for rains of the same intensity but different origins. The drop-size characteristics of convectional and frontal rain are different as are those of rain formed at the warm and cold fronts of a temperate depression.

In spite of the difficulties posed by these variations it is possible to derive a general relationship between kinetic energy and rainfall intensity. Based on the work of Laws and Parsons (1943), Wischmeier and Smith (1958) obtained the equation:

$$KE = 13.32 + 9.78 \log_{10} I, \tag{3.1}$$

where I is the rainfall intensity (mm h^{-1}) and KE is the kinetic energy (J m^{-2} mm^{-1}). For tropical rainfall, Hudson (1965) gives the equation:

$$KE = 29.8 - \frac{127.5}{I} \tag{3.2}$$

To compute the kinetic energy of a storm, a trace of the rainfall from an automatically recording rain gauge is analysed and the storm divided into small time increments of uniform intensity. For each time period, knowing the intensity of the rain, the kinetic energy of rain at that intensity is estimated from one of the above equations and this, multiplied by the amount of rain received, gives the kinetic energy for that time period. The sum of the kinetic energy values for all the time periods gives the total kinetic energy of the storm (Table 3.4).

Table 3.4 Calculation of erosivity

TIME FROM START (min)	RAINFALL (mm)	INTENSITY (mm h^{-1})	KINETIC ENERGY (J m^{-2} mm^{-1})	TOTAL KINETIC ENERGY (col 2 × col 4) (J m^{-2})
0–14	1.52	6.08	8.83	13.42
15–29	14.22	56.88	27.56	391.90
30–44	26.16	104.64	28.58	747.65
45–59	31.50	126.00	28.79	906.89
60–74	8.38	33.52	26.00	217.88
75–89	0.25	1.00	—	—

Kinetic energy is calculated from equation (3.2).

Erosivity indices

Wischmeier index (EI_{30})

maximum 30-min rainfall   =   26.16+31.50

  =   57.66mm

maximum 30-min intensity =   57.66×2

  =   115.32mm h^{-1}

total kinetic energy   =   total of column 5

  =   2 277.74 J m^{-2}

EI_{30} =   2 277.74×115.32

  =   262 668.98 J m^{-2} .mm h^{-1}

Hudson index (KE >25)

total kinetic energy for   =   total of column 5, lines 2, 3, 4 and 5 only
rainfall intensity⩾25mm h^{-1}

  =   2 264.32 J m^{-2}

To be valid as an index of potential erosion, an erosivity index must be significantly correlated with soil loss. Wischmeier and Smith (1958) found that soil loss by splash, overland flow and rill erosion is related to a compound index of kinetic energy and the maximum 30-min rainfall intensity (I_{30}). This index, known as EI_{30}, is open to criticism. First, being based on estimates of kinetic energy using equation (3.1), it is of suspect validity for tropical rains of high intensity. Second, it assumes that erosion occurs even with light intensity rain whereas Hudson (1965) has shown that erosion is almost entirely caused by rain falling at intensities greater than 25 mm h^{-1}. The inclusion of I_{30} in the index is an attempt to correct for overestimating the importance of light intensity rain but it is not entirely successful because the ratio of intense erosive rain to non-erosive rain is not well correlated with I_{30} (Hudson, personal communication). In fact, there is no obvious reason why the maximum 30-min intensity is the most appropriate parameter to choose. Stocking and Elwell (1973a) recommend its use only for bare soil conditions. With sparse and dense plant covers they obtain better correlations with soil loss using the maximum 15- and 5-min intensities respectively. A different approach is suggested from the research of Vuillaume (1969) at Kountkouzout, Niger, where the character of the rain in the first twenty minutes of a storm appears to be the most critical factor. It is during this period that the soil becomes saturated and the most intense rains occur. The latter is unlikely to be the case world-wide, however, particularly in temperate latitudes subjected to prolonged rains of low intensity.

As an alternative erosivity index, Hudson (1965) uses KE > 25 which, to compute for a single storm, means summing the kinetic energy received in those time increments when the rainfall intensity equals 25 mm h^{-1} or greater (Table 3.4). When applied to data from Rhodesia, a better correlation was obtained between this index and soil loss than between soil loss and EI_{30}. Stocking and Elwell (1973a) have reworked Hudson's data and, incorporating more recent information, have suggested that EI_{30} is the better index after all. Since they compute EI_{30} only for storms yielding 12.5 mm of rain and with a maximum 5-min intensity greater than 25 mm h^{-1}, they have removed most of the objections to the original EI_{30} index, however, and produced an index which is philosophically very close to KE > 25. Hudson's index has the advantages of simplicity and less stringent data requirements. Although somewhat limiting for temperate latitudes, it can be modified by using a lower threshold value such as KE > 10 (Morgan, 1977).

By calculating erosivity values for individual storms over a period of 20 to 25 years, mean monthly and mean annual data can be obtained. Unfortunately, the EI_{30} and KE > 25 indices yield vastly different values because of the inclusion of I_{30} in the former. Rapp, Axelsson, Berry and Murray-Rust (1972) quote figures for Morningside Farm in the Morogoro region of Tanzania of 70 000 for the annual value of EI_{30} and 12 000 for KE > 25. The two indices cannot be substituted for each other.

3.1.3. Wind erosivity

Few studies have been made of the return periods of different wind velocities but a simple index of erosivity as a function of the velocity and duration of the wind has been developed by Skidmore and Woodruff (1968). The erosivity of wind blowing in vector j is obtained from

$$EW_j = \sum_{i=1}^{n} \overline{Vt}^3_{ij} \quad f_{ij},\tag{3.3}$$

where EW_j is the wind erosivity value for vector j, $\overline{Vt}$ is the mean velocity of wind in the ith speed group for vector j above a threshold velocity taken as 19 km h^{-1}, and f_i is the duration of the wind for the vector j in the ith speed group. Expanding this equation for total wind erosivity (EW) over all vectors yields:

$$EW = \sum_{j=0}^{15} \sum_{i=1}^{n} \overline{Vt^3}_{ij} \quad f_{ij},$$ (3.4)

where vectors $j = 0$ to 15 represent the sixteen principal compass directions beginning with $j = 0 = E$ and working anticlockwise so that $j = 1 = ENE$ and so on.

3.2. ERODIBILITY

Erodibility defines the resistance of the soil to both detachment and transport. Although soil resistance to erosion depends in part on topographic position, slope steepness and the amount of disturbance created by man, for example during tillage, the properties of the soil are the most important determinants. Erodibility varies with soil texture, aggregate stability, shear strength, infiltration capacity and organic and chemical content.

The role of soil texture has been indicated in Chapter 2 where it was shown that large particles are resistant to transport because of the greater force required to entrain them and that fine particles are resistant to detachment because of their cohesiveness. The least resistant particles are silts and fine sands. Thus soils with a high silt content are erodible. Richter and Negendank (1977) show that soils with 40 to 60 per cent silt content are the most erodible. Evans (in press) prefers to examine erodibility in terms of the clay content, indicating that soils with a restricted clay fraction, between 9 and 30 per cent, are most susceptible to erosion.

The use of the clay content as an indicator of erodibility is theoretically more satisfying because the clay particles combine with organic matter to form soil aggregates or clods and it is the stability of these which determines the resistance of the soil. Soils with a high content of base minerals are generally more stable as these contribute to the chemical bonding of the aggregates. Stability also depends on the type of clay mineral present. Illite and montmorillonite more readily form aggregates but the more open lattice structure of these minerals and the greater swelling and shrinkage which occurs on wetting and drying render the aggregates less stable than those formed from kaolinite.

The shear strength of the soil is a measure of its cohesiveness and resistance to shearing forces. Although this is relevant to a limited extent to the response of the soil to the impact of running water and wind, it is more useful as an indicator of potential mass movement (Bryan, 1968).

Infiltration capacity, the maximum sustained rate at which soil can absorb water, is influenced by pore size, pore stability and the form of the soil profile. Soils with stable aggregates maintain their pore spaces better whilst soils with swelling clays or minerals that are unstable in water tend to have low infiltration capacities. Although estimates of the infiltration capacity can be obtained in the field using infiltrometers (Hills, 1970), actual capacities during storms are often much less than those indicated by field tests. Infiltration capacities of the Oligocene Tongrian Sands in Belgium are in excess of 200 mm h^{-1} according to field measurements but runoff can occur with rains of only 20 mm h^{-1} (De Ploey, 1977). Similar discrepancies between measured infiltration capacities and rainfall intensities resulting in runoff have been observed on soils of the Lower Greensand in Bedfordshire (Morgan, 1977). Where soil properties vary with profile depth, it is the horizon with the lowest infiltration capacity which is critical. In the case of these sandy soils, the critical horizon is often the surface where a crust of 2 mm thickness may be sufficient to decrease infiltration capacity enough to cause runoff, even though the underlying soil may be dry. A further factor which influences infiltration capacity is rainfall intensity and there is evidence to suggest that, instead of being a constant value, infiltration capacity increases with rainfall intensity (Nassif and Wilson, 1975; De Ploey, Savat and Moeyersons, 1976). Thus, increasing intensity of rain may not lead to a corresponding increase in runoff and decreasing intensity may even lead to runoff.

The organic and chemical constituents of the soil are important because of their influence on aggregate stability. Soils with less than 2 per cent organic matter can be considered erodible (Evans, in press).

Many attempts have been made to devise a simple index of erodibility based either on the properties of the soil as determined in the laboratory or the field, or on the

Table 3.5 Indices of soil erodibility

STATIC LABORATORY TESTS

Dispersion ratio	Comparing silt+clay content of undispersed soil with that of soil dispersed in water	Middleton (1930)
Clay ratio	$\dfrac{\text{Sand}+\text{silt \%}}{\text{clay \%}}$	Bouyoucos (1935)
Surface aggregation ratio	$\dfrac{\text{surface area of particles} >0.05\text{mm}}{\substack{\text{\% silt}+\text{clay in} \\ \text{dispersed soil}-}}$ % silt+clay in undispersed soil	André and Anderson (1961)

STATIC FIELD TESTS

Erodibility index	$\dfrac{1}{\text{mean shearing resistance} \times \text{permeability}}$	Chorley (1959)

DYNAMIC LABORATORY TESTS

Simulated rainfall test	Comparing erosion of different soils subjected to a standard storm	Woodburn and Kozachyn (1956)
Water-stable aggregate content	% of water stable aggregates >0.5mm	Bryan (1968)
Erosion index (E)	$\dfrac{\text{index of dispersion} \times \text{index of water-retaining capacity}}{\text{index of aggregation}}$	Voznesensky and Artsruui (1940)

DYNAMIC FIELD TESTS

Erosion index (K)	soil loss per unit of EI_{30}	Wischmeier and Mannering (1969)

response of the soil to rainfall and wind (Table 3.5). In a review of the indices applied to water erosion, Bryan (1968) favours aggregate stability as the most efficient index. He uses the proportion of water-stable non-primary aggregates larger than 0.5 mm contained in the soil as an indicator of erodibility, the greater the proportion, the more resistant being the soil to erosion.

A more commonly used index is the K value which represents the soil loss per unit of EI_{30}, as measured in the field on a standard bare soil plot, 22 m long and of 5° slope. Estimates of the K value may be made if the grain-size distribution, organic content, structure and permeability of the soil are known (Wischmeier, Johnson and Cross, 1971; Fig. 3.2).

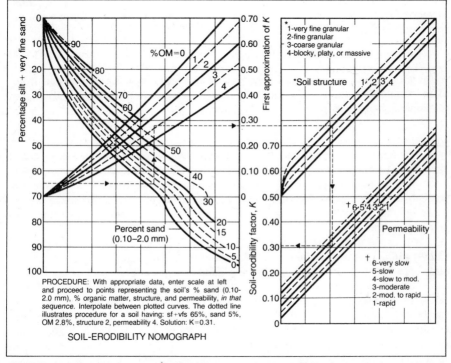

SOIL-ERODIBILITY NOMOGRAPH

PROCEDURE: With appropriate data, enter scale at left and proceed to points representing the soil's % sand (0.10-2.0 mm), % organic matter, structure, and permeability, *in that sequence*. Interpolate between plotted curves. The dotted line illustrates procedure for a soil having: sf+vfs 65%, sand 5%, OM 2.8%, structure 2, permeability 4. Solution: K=0.31.

Fig. 3.2 Nomograph for computing the K value of soil erodibility for use in the Universal Soil-Loss Equation (after Wischmeier, Johnson and Cross, 1971).

The resistance of the soil to wind erosion depends upon dry rather than wet aggregate stability and on the moisture content, wet soil being less erodible than dry soil, but is otherwise related to much the same properties as affect its resistance to water erosion. Chepil (1950), using wind tunnel experiments, has related wind erodibility of soils to various indices of dry aggregate structure but little of the work has been tested in field conditions. The wind resistance of soils can also be expressed by an index related to cohesion. Taking a value of 100 to represent maximum cohesion, Shiyatyy, Lavrovskiy and Khmolenko (1972) derive the following index of soil resistance (R):

$$R = 100 - C,\qquad(3.5)$$

where C is a measure of actual cohesion based on the clay and sand fractions of the soil, and defined by

$$C = 34.7 + 0.9X_1 - 0.3X_2 - 0.4X_3, \tag{3.6}$$

where X_1, X_2 and X_3 denote the proportions of the soil falling in the textural classes < 0.001 mm, 0.05–0.25 mm and > 0.25 mm respectively.

3.3. EFFECT OF SLOPE

Erosion would normally be expected to increase with increases in slope steepness and slope length as a result of respective increases in velocity and volume of surface runoff. Further, whilst on a flat surface raindrops splash soil particles randomly in all directions, on sloping ground more soil is splashed downslope than upslope, the proportion increasing as the slope steepens. The relationship between erosion and slope can be expressed by the equation:

$$Qs \propto \tan^m \theta \, L^n \tag{3.7}$$

where θ is the gradient angle and L is slope length. Zingg (1940), in a study of data from five experimental stations of the United States Soil Conservation Service, found that the relationship had the form:

$$Qs \propto \tan^{1.4} L^{0.6} \tag{3.8}$$

where Qs is expressed per unit area. To express Qs proportional to distance downslope, the value of n must be increased by 1.0. Since the values for the exponents have been confirmed in respect of m by Musgrave (1947) and m and n by Kirkby (1969b) there is some evidence to suggest that equation (3.8) has general validity. Other studies show, however, that the values are sensitive to the interaction of other factors in the erosion–slope relationship.

Working with data from experimental stations in Rhodesia, Hudson and Jackson (1959) found that m was close to 2.0 in value, indicating that the effect of slope is stronger under tropical conditions where rainfall is heavier. The effect of soil is illustrated by the laboratory experiments of Gabriels, Pauwels and De Boodt (1975). These authors show that the value of m increases with the grain size of the material, from 0.6 for particles of 0.05 mm to 1.7 for particles of 1.0 mm. The value of m may also be sensitive to slope itself, decreasing with increasing slope steepness. Thus, Horváth and Erödi (1962) found from their laboratory studies that $m = 1.6$ for slopes between 0° and 2.5°, 0.7 for slopes between 3° and 6.5°, and 0.4 for slopes over 6.5°. On still steeper slopes the value may be expected to decrease further as soil-covered slopes give way to rock surfaces and soil supply becomes a limiting factor. Heusch (1970) obtained a value of -3.8 for slopes of 6.5° to 33° in Morocco, attributing the negative exponent to decreasing surface runoff and increasing subsurface flow on the steeper slopes. The exponents vary in value with slope shape. Examining soil loss from 3 m long plots on slopes of average steepness between 2° and 8° under simulated rainfall, D'Souza and Morgan (1976) obtained values for m of 0.5 for convex slopes, 0.4 for straight slopes and 0.14 for concave slopes.

Few studies have examined the effect of varying plant cover and vegetation densities on the exponent values. Quinn, Morgan and Smith (in preparation investigated the change in the value of m for soil loss from 1.2 m long plots with slopes of 5° to 30° under simulated rainfall in relation to decreasing grass cover brought about by trampling. They found that $m = 0.7$ for fully grassed straight slopes, rises to 1.9 in the early stages of trampling before falling to 1.1 when only about 25 per cent of the grass cover remains.

All the values discussed thus far relate to erosion by overland flow. Kirkby (1969*b*; 1971) notes variations in the values with the processes of erosion. Values of n are zero for soil creep and splash erosion, range between 0.3 and 0.7 for erosion by overland flow and rise to between 1.0 and 2.0 if rilling occurs. Values of m are 1.0 for soil creep, range between 1.0 and 2.0 for splash and between 1.3 and 2.0 for overland flow, and may be as high as 3.0 for rivers. Meyer and Wischmeier (1969) assume that $m = 1.0$ for rainsplash whilst laboratory studies by Moeyersons and De Ploey (1976) yield $m = 0.75$. Limited evidence from field studies in Bedfordshire (Morgan, 1978) suggests that for splash, m may be closer to 2.0 in value.

3.4. EFFECT OF PLANT COVER

The importance of a plant cover in reducing erosion is demonstrated by experiments at the Henderson Research Station in Rhodesia (Hudson, 1971) where, in the period 1953–56, mean annual soil loss from bare ground was 4.63 kg m^{-2} compared with 0.04 kg m^{-2} from ground with a dense cover of *Digitaria*. The major role of vegetation is in interception of the raindrops so that their kinetic energy is dissipated by the plants rather than imparted to the soil. This role is emphasized by the mosquito gauze experiment of Hudson and Jackson (1959) in which soil loss was compared from two identical bare soil plots. Over one plot was suspended a fine wire gauze which had the effect of breaking the force of the raindrops, absorbing their impact and allowing the water to fall to the ground from a low height as a fine spray. The mean annual soil loss over a six year period was 141.3 m^3 ha^{-1} for the open plot and 1.2 m^3 ha^{-1} for the plot covered by gauze.

The effectiveness of a plant cover in reducing erosion depends upon the height and continuity of the canopy, the density of the ground cover and the root density. The height of the canopy is important because water drops falling from 7 m may attain over 90 per cent of their terminal velocity. Further, raindrops intercepted by the canopy may coalesce on the leaves to form larger drops which are more erosive. A ground cover not only intercepts the rain but also dissipates the energy of running water and wind, imparts roughness to the flow and thereby reduces its velocity. Since erosion rates vary with either the cube or the fifth power of velocity, the effect on soil loss is considerable. The main effect of the root network is in opening up the soil, thereby enabling water to penetrate and increasing infiltration capacity.

Generally, forests are the most effective in reducing erosion because of their canopy but a dense growth of grass may be almost as efficient. For adequate erosion protection at least 70 per cent of the ground surface must be covered (Fournier, 1972; Elwell and Stocking, 1976).

Many of the relationships examined in this chapter have been only empirically established and, as is common with empirical studies, there is much disagreement over the precise nature of the equations which express the effect of the various controlling factors on soil loss. At the present time detailed process studies in the field and the laboratory are increasing our understanding of the relationships. Until a sufficient number of these studies has been carried out, the somewhat equivocal evidence of the way in which the soil erosion system operates must provide the foundation for evaluating the risk of excessive soil loss.

CHAPTER 4
EROSION HAZARD
ASSESSMENT

The assessment of erosion hazard is a specialized form of land resource evaluation, the objective of which is to identify those areas of land where the maximum sustained productivity from a given landuse is threatened by excessive soil loss. The assessment aims at dividing a land area into regions, similar in their degree and kind of erosion hazard, as a basis for planning soil conservation work. Since erosion risk is closely related to the plant cover and therefore to the use made of the land, soil erosion surveys often form part of a broader land resources study. In the same way that, depending on their objective, resource studies are undertaken at reconnaissance, semi-detailed and detailed levels, so erosion surveys may be carried out at several scales. In this chapter, generalized and semi-detailed assessments are examined.

4.1. GENERALIZED ASSESSMENTS

Generalized assessments of erosion risk are made at the macro-, often national, scale, and, as befits surveys at this scale (Table 1.1), are based largely on analysing climatic data.

4.1.1. Using erosivity indices

Erosivity data may be used as an indicator of regional variations in erosion potential to pinpoint areas of high risk. Stocking and Elwell (1976) present a generalized picture of erosion risk in Rhodesia, based on mean annual erosivity values, showing that high-risk areas are in the Eastern Districts, the region east of Fort Victoria, and the High and Middle Veld north and east of Salisbury (Fig. 4.1). The area south and west of Bulawayo, by contrast, has a much lower risk. Temporal variations in erosion risk are revealed by the mean monthly erosivity values. Hudson (1971; Fig. 4.2) contrasts the erosivity patterns of Bulawayo and Salisbury. At Bulawayo, erosivity is low at the beginning of the wet season and increases as the season progresses. By the time the maximum values are experienced the plant cover has had a chance to become established, giving protection against erosion and lowering the risk. At Salisbury, however, erosivity is highest at the time of minimal vegetation cover.

In many countries insufficient rainfall records from autographic gauges are available to calculate erosivity nationwide. In these cases, an attempt is made, for the recording stations where erosivity can be determined, to find a more widely available rainfall parameter which significantly correlates with erosivity and from which erosivity values may be predicted using a least-squares regression equation. Roose (1975),

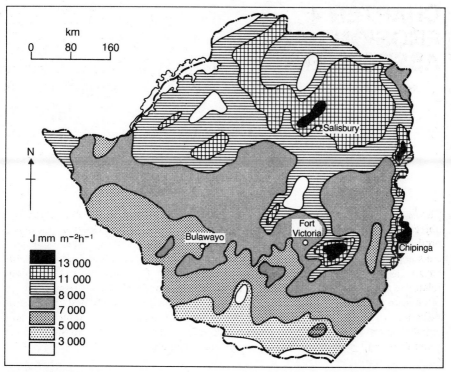

Fig. 4.1 Mean annual erosivity in Rhodesia (after Stocking and Elwell, 1976).

examining data in the Ivory Coast and Upper Volta, finds that mean annual EI_{30} values can be approximated by the mean annual rainfall totals (mm) multiplied by 50. Ateshian (1974) built up a map of erosivity west of the Rocky Mountains based on a power function relating mean annual erosivity to the 6 h rainfall total with a two-year return period.

4.1.2. Rainfall aggressiveness

The most commonly used index, which has been shown to be significantly correlated with sediment yields in rivers (Fournier, 1960), is the ratio p^2/P, where p is the highest mean monthly precipitation and P is the mean annual precipitation. It is strictly an index of the concentration of precipitation into a single month and thereby gives a crude measure of the intensity of the rainfall and, in so far as a high value denotes a strongly seasonal climatic regime with a dry season during which the plant cover decays, of erosion protection by vegetation. Using data from seventy-eight drainage basins, Fournier (1960) derived the following empirical relationship between mean annual sediment yield (Qs; g m^{-2}), mean altitude (H; m) and mean slope of the basin (S):

$$\log Qs = 2.65 \log \frac{p^2}{P} + 0.46 (\log H)(\tan S) - 1.56 \qquad (4.1)$$

This equation has been used by Low (1967) to investigate regional variations in erosion risk in Peru.

It was stated in Chapter 3 that soil loss varies with both the intensity and the amount of rain. In the absence of data from which to estimate the kinetic energy of rainfall, an

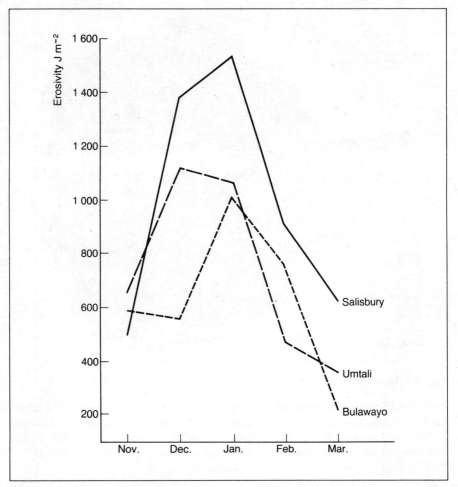

Fig. 4.2 Mean monthly erosivity for three towns in Rhodesia (after Hudson, 1965).

index incorporating these two parameters should be useful. Lal (1976) combines them in the index AI_m, where A is the total rainfall (cm) and I_m is the maximum rainfall intensity (cm h^{-1}). Comparing this index with EI_{30} and KE > 25, he finds that the correlation with runoff and soil loss from field plots at Ibadan, Nigeria is better, though, since AI_m and EI_{30}, in their respective units of cm^2 h^{-1} and J mm m^{-2} h^{-1}, give identical values, it is not clear why this should be. The major advantage of AI_m would appear to be its ease of calculation.

4.1.3. Factorial scoring

A simple scoring system for rating erosion risk has been devised by Stocking and Elwell (1973b) in Rhodesia. Taking a 1:1 000 000 base map, the country is divided on a grid system into units of 184 km^2. Each unit is rated on a scale from 1 to 5 in respect of erosivity, erodibility, slope, ground cover and human occupation, the latter taking account of the density and the type of settlement. The scoring is arranged so that 1 is associated with a low risk of erosion and 5 with a high risk. The five factor scores are

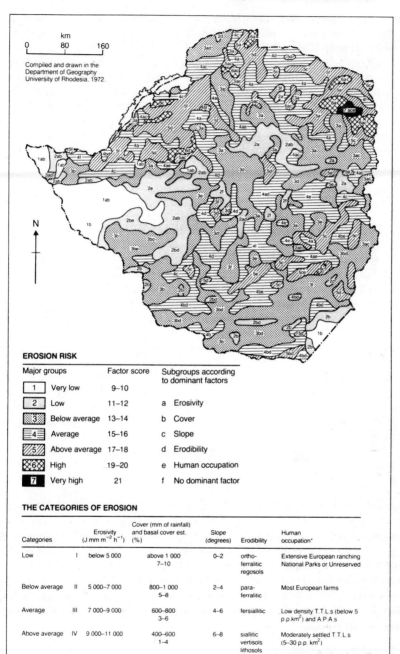

EROSION RISK

Major groups		Factor score	Subgroups according to dominant factors
1	Very low	9–10	
2	Low	11–12	a Erosivity
3	Below average	13–14	b Cover
4	Average	15–16	c Slope
5	Above average	17–18	d Erodibility
6	High	19–20	e Human occupation
7	Very high	21	f No dominant factor

THE CATEGORIES OF EROSION

Categories		Erosivity (J mm m^{-2} h^{-1})	Cover (mm of rainfall) and basal cover est. (%)	Slope (degrees)	Erodibility	Human occupation*
Low	I	below 5 000	above 1 000 7–10	0–2	ortho-ferralitic regosols	Extensive European ranching National Parks or Unreserved
Below average	II	5 000–7 000	800–1 000 5–8	2–4	para-ferralitic	Most European farms
Average	III	7 000–9 000	600–800 3–6	4–6	fersiallitic	Low density T.T.L.s (below 5 p.p.km^2) and A.P.A.s
Above average	IV	9 000–11 000	400–600 1–4	6–8	siallitic vertisols lithosols	Moderately settled T.T.L.s (5–30 p.p. km^2)
High	V	above 11 000	below 400 0–2	above 8	non-calcic hydro-morphic sodic	Densely-settled T.T.L.s (above 30 p.p. km^2)

(*Notes:* Cover. Erodibility. and Human occupation are only tentative and cannot as yet be expressed on a firm quantitative basis)
*p.p. km^2 — persons per square kilometre.   T.T.L.–Tribal Trust Land.   A.P.A.–African Purchase Areas.

Fig. 4.3 Erosion survey of Rhodesia (after Stocking and Elwell, 1973*b*).

summed to give a total score which is compared with an arbitrarily chosen classification system to categorize areas of low, moderate and high erosion risk. The scores are mapped and areas of similar risk delineated (Fig. 4.3).

Several problems are associated with this technique. First, the classification may be sensitive to different scoring systems. For example, the use of different slope groups may yield different assessments of the degree of erosion risk. Second, each factor is treated independently whereas, as shown in Chapter 3, there is interaction between the factors. Slope steepness may be much more important in areas of high than in areas of low erosivity. Third, the factors are combined by addition. There is no reason why this should be a more appropriate method of combining them than multiplication. Fourth, each factor is given equal weight. Despite these difficulties, the technique is easy to use and has the advantage that factors which cannot be easily quantified in any other way can be readily included.

4.2. SEMI-DETAILED ASSESSMENT

4.2.1. Land capability classification

The land capability classification was developed by the United States Soil Conservation Service as a method of assessing the extent to which limitations such as erosion risk, soil depth, wetness and climate hinder the agricultural use that can be made of the land. The United States classification (Klingebiel and Montgomery, 1966) has been adapted for use in many other countries (Hudson, 1971).

The objective of the classification is to regionalize an area of land into units with similar kinds and degrees of limitation. The basic unit is the capability unit. This consists of a group of soil types of sufficiently similar conditions of profile form, slope and degree of erosion as to make them suitable for similar crops and warrant the use of similar conservation measures. The capability units are combined into sub-classes according to the nature of the limiting factor and these, in turn, are grouped into classes based on the degree of limitation. The United States system recognizes eight classes arranged from Class I, characterized by no or very slight risk of damage to the land when used for cultivation, to Class VIII, very rough land which can be safely used only for wildlife, limited recreation, and watershed conservation. The first four classes are designated as suitable for arable farming (Table 4.1). Assigning a tract of land to its appropriate class is aided by the use of a flow diagram (Fig. 4.4).

Although the inclusion of many soil properties in the classification may seem to render it useful for landuse planning generally, it must be appreciated that, as befits a classification evolved in the wake of the erosion scare in the United States in the 1930s, its bias is towards soil conservation. This bias is illustrated by the dominance of slope as a factor in determining capability class and by the emphasis given to soil conservation in the recommendations on how each class of land should be treated. Attempts to use the classification in a wider sphere have only drawn attention to its limitations. The classification does not specify the suitability of the land for particular crops. A separate land suitability classification has been devised to do this (Vink, 1975). The assigning of a capability class is not an indicator of land value which may reflect the scarcity of a certain type of land, nor is it a measure of whether the farmer can make a profit, which is much influenced by the market prices of the crops grown and the farmer's skill. The stress laid by the classification on arable farming can also be a disadvantage. In Malaysia, where it conflicts with the priorities set for national development, the Economic Planning Unit has modified the system by defining Class I land as that with potential for mineral development (Panton, 1969; Table 4.2). Insufficient attention is given to the recrea-

Table 4.1   Land capability classes (United States system)

CLASS	CHARACTERISTICS AND RECOMMENDED LANDUSE
I	Deep, productive soils, easily worked, on nearly level land; not subject to overland flow; no or slight risk of damage when cultivated; use of fertilizers and lime, cover crops, crop rotations required to maintain soil fertility and soil structure.
II	Productive soils on gentle slopes; moderate depth; subject to occasional overland flow; may require drainage; moderate risk of damage when cultivated; use crop rotations, water-control systems or special tillage practices to control erosion.
III	Soils of moderate fertility on moderately steep slopes, subject to more severe erosion; subject to severe risk of damage but can be used for crops provided adequate plant cover is maintained; hay or other sod crops should be grown instead of row crops.
IV	Good soils on steep slopes, subject to severe erosion; very severe risk of damage but may be cultivated occasionally if handled with great care; keep in hay or pasture but a grain crop may be grown once in five or six years.
V	Land is too wet or stony for cultivation but of nearly level slope; subject to only slight erosion if properly managed; should be used for pasture or forestry but grazing should be regulated to prevent plant cover from being destroyed.
VI	Shallow soils on steep slopes; use for grazing and forestry; grazing should be regulated to preserve plant cover; if the plant cover is destroyed, use should be restricted until cover is re-established.
VII	Steep, rough, eroded land with shallow soils; also includes droughty and swampy land; severe risk of damage even when used for pasture or forestry; strict grazing or forest management must be applied.
VIII	Very rough land; not suitable even for woodland or grazing; reserve for wildlife, recreation or watershed conservation.

Classes I–IV denote soils suitable for cultivation.
Classes V–VIII denote soils unsuitable for cultivation.
(Modified from Stallings, 1957).

tional use of land. The capability classification implies that land is set aside for recreation only when it is too marginal for arable or pastoral farming but, as shown by studies of footpath deterioration in upland areas (Bayfield, 1973), such land is often marginal for recreational use too. This fact highlights the difficulty of incorporating agricultural and non-agricultural activities in a single classification. One approach to this problem is provided by the Canada Landuse Inventory (McCormack, 1971) which employs four separate classifications covering agriculture, forestry, recreation and wildlife.

As a result of these criticisms, the land capability classification has tended, in recent years, to be discredited, often unfairly, as many of the criticisms arise from attempts to use the classification for purposes for which it was not intended. The value of land

LAND CAPABILITY CLASS	I	II		III		IV	
Permissible slope	0°–1°	0°–1°	1°–2.5°	0°–2.5°	2.5°–4.5°	0°–4.5°	4.5°–7°
Minimum effective depth (Texture here refers to average textures)	1 m of CL or heavier	50 cm of Sal. or heavier	50 cm of SaCl or heavier	50 cm of S or LS 25cm of Sal. or heavier	(a) 50 cm of SaCL (b) 25 cm of CL or heavier	25 cm of any texture	25 cm of SaCl or heavier
Texture of surface soil	CL or heavier	Sal or heavier S, or LS if upper subsoil is Sal or heavier	Sal or heavier	No direct limitations	(a) Sal or heavier (b) CL or heavier	No direct limitations	Sal or heavier
Permeability 5 or 4 to at least–	1 m	50 cm	50 cm	No direct limitations	No direct limitations	No direct limitations	
Not worse than 3 to–		1 m	1 m	1 m or 50 cm if average texture is CL is heavier	1 m		
Physical characteristics of the surface soil– Permissible symbols	Not permitted	t1	t1	t1 and t2	t1 and t2	t1 and t2	t1 and t2
Erosion– Permissible symbols	1	1 and 2	1 and 2	1,2 and 3	1, 2 and 3	1, 2 and 3	1, 2 and 3
Wetness criteria– Permissible symbols	Not permitted	w1	w1	w1	w1	w1 and w2	w1 and w2

S = Sand
Sal = Sandy Loam
LS = Loamy Sand
CL = Clay Loam
SaCL = Sandy Clay Loam

Fig. 4.4 Criteria and flow chart for determining land capability class according to the Classification of the Federal Government, Rhodesia and Nyasaland (after Hudson, 1971).

capability assessment lies in identifying the risks attached to cultivating the land and in indicating the soil conservation measures which are required. Improvements to the classification rest on making the conservation recommendations more specific, as is the case with the treatment-oriented scheme developed in Taiwan and tested in hilly land in Jamaica by Sheng (1972a; Table 4.3; Fig. 4.5).

4.2.2. Land systems classification

Land systems analysis is used to compile information on the physical environment for the purpose of resource evaluation. The land is classified into areal units, termed land systems, which are made up of smaller units, land facets, arranged in a clearly recurring pattern. Since land facets are defined by their uniformity of landform, especially slope, soils and plant community, land systems comprise an assemblage of landform, soil and vegetation types. Land systems classification has been described and discussed in many texts (Cooke and Doornkamp, 1974; Young, 1976) as a method of integrated resource

NOTES

Effective depth (m)

1	Deep	More than 1.5 m
2	Moderately deep	1 m to 1.5 m
3	Moderately shallow	50 cm to 1 m
4	Shallow	25 cm to 50 cm
5	Very shallow	Less than 25 cm

Texture of surface soil

A	Sand	More than 85% sand
X	Loamy sand	80–85% sand
B	Sandy loam	Less than 20% clay; 50–80% sand
C	Sandy clay loam	20–30% clay; 50–80% sand
D	Clay loam	20–30% clay; less than 50% sand
E	Sandy clay	More than 30% clay; 50–70% sand
F	Clay	30–50% clay; less than 50% sand
G	Heavy clay	More than 50% clay

Permeability	Description	Rate of flow*
	Very slow	Less than 1.25
	Slow	1.25 to 5
	Moderately slow	5 to 20
	Moderate	20 to 65
	Moderately rapid	65 to 125
	Rapid	125 to 250
	Very rapid	over 250

* The rate of flow in mm per hour through saturated undisturbed cores under a head of 12.5 mm of water.

Physical characteristics of surface soil

t1	Slightly unfavourable physical conditions. The soil has a tendency to compact and seal at the surface and a good tilth is not easily obtained.
t2	Unfavourable physical conditions. Compaction and sealing of the surface soil are more severe. A hard crust forms when the bare soil is exposed to rain and sun and poor emergence of seedlings can severely reduce the crop. On ploughing, large clods are turned up which are not easily broken.

Erosion

1	No apparent, or slight, erosion.
2	Moderate erosion : moderate loss of topsoil generally and/or some dissection by run-off channels or gullies.
3	Severe erosion, severe loss of topsoil generally and/or marked dissection by run-off channels or gullies.
4	Very severe erosion: complete truncation of the soil profile and exposure of the subsoil (B horizon) and/or deep and intricate dissection by run-off channels or gullies.

Wetness

w1	Wet for relatively short and infrequent periods.
w2	Frequently wet for considerable periods.
w3	Very wet for most of the season.

Fig. 4.4 (cont.)

Fig. 4.4 (*cont.*)

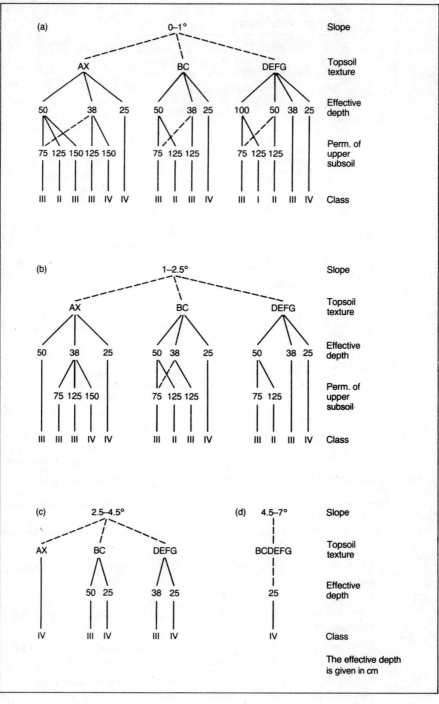

Fig. 4.4 (*cont.*)

ADDITIONAL REQUIREMENTS

*Factors affecting cultivation

g	Gravelly or stony
b Downgrade Class I to II	Very gravelly or stony
o	Bouldery
s	Very bouldery
v Class VI	Outcrops
r	Extensive outcrops

*Permeability

75 to 500–
otherwise class IV
Not applicable to
basalts or norites

*Erosion

Class: I : 1
        II : 1; 2
        III : 1; 2; 3

*'t' Factors

Class: II : t1
        III : t1; t2
        IV : t1; t2

*Wetness

Class: II : w1
        III : w1
        IV : w2
        V : w3

w2 downgrades Class II and III to IVw unless the land is already
Class IV on code, in which case it remains as Class IV.

*Note

Any land not meeting the minimum requirements shown
on this sheet is Class VI.

Table 4.2   Land capability classes recognized by the Economic Planning Unit,
Malaysia

CLASS	DESCRIPTION
I	Land with a high potential for mineral development.
II	Land with a high potential for agricultural development with a wide range of crops.
III	Land with a moderate potential for agricultural development with a restricted range of crops; best used for crops with a wide range of soil tolerance.
IV	Land with potential for productive forest development; best suited to commercial timber exploitation.
V	Land with little or no mineral, agricultural or forest development potential; suitable for development as protective reserves for conservation, water catchment, game, recreation, or similar purpose.

After Panton (1969).

Table 4.3 Treatment-oriented land capability classification

GROUP	CLASS	CHARACTERISTICS AND RECOMMENDED TREATMENTS
Suitable for tillage	C_1	Up to 7° slope; soil depth normally over 10cm; contour cultivation; strip cropping; broadbase terraces.
	C_2	Slopes 7°–15°; soil depth over 20cm; bench terracing (construction by bulldozers); use of four-wheel tractors.
	C_3	Slopes 15°–20°; soil depth over 20 cm; bench terracing on deep soil (construction by small machines); silt-pits on shallower soils; use of small tractors or walking tractors.
	C_4	Slopes 20°–25°; soil depth over 50 cm; bench terracing and farming operations by hand labour.
	P	Slopes 0°–25°; soil depth too shallow for cultivation; use for improved pasture on rotational grazing system; system; zero grazing where land is wet.
	FT	Slopes 25°–30°; soil depth over 50 cm; use for tree crops with bench terracing; inter-terraced areas in permanent grass; use contour planting; diversion ditches; mulching.
	F	Slopes over 30° or over 25° where soil is too shallow for tree crops; maintain as forest land.
Wetland, liable to flood; also stony land	P	Slopes 0°–25°; use as pasture.
	F	Slopes over 25°; use as forest.
Gullied land	F	Maintain as forest land.

After Sheng (1972a).
The scheme is most suitable for hilly lands in the tropics.

survey, a role which lies beyond the scope of this book. What is important here, considering that many of the factors examined in land systems analysis are relevant to the soil erosion system, is its value for erosion risk evaluation. The extent to which land systems conform to discrete units of kind and degree of erosion hazard was examined by Higginson (1973) in the Hunter Valley, New South Wales. He carried out an erosion survey of the area from aerial photographs, using a classification recognizing seven grades of erosion severity, and analysed the distribution of erosion classes falling within each land system. Far from each system being distinct, he found that the land systems could be combined into four groups. It would seem, therefore, that the relationship between erosion risk and land systems is poor and that the land systems classification provides only a generalized assessment. To a large extent this is expected because, by their make-up, land systems are likely to comprise areas of markedly different erosion risk. A similar study at the land facet level might prove more fruitful.

4.2.3. Soil erosion survey

The first types of erosion survey were essentially static in concept and consisted of mapping, often from aerial photographs, the rills, gullies and blow-outs occurring in the area (Jones and Keech, 1966; Vink, 1968). Erosion hazard was estimated by calculating

Slope / Soil Depth	1. Gentle sloping < 7°	2. Moderate sloping 7° – 15°	3. Strongly sloping 15° – 20°	4. Very strongly sloping 20° – 25°	5. Steep 25° – 30°	6. Very steep > 30°
Deep (D) > 36 in. (> 90 cm)	C_1	C_2	C_3	C_4	FT	F
Moderately deep (MD) 20 – 36 in. (50 – 90 cm.)	C_1	C_2	C_3	C_4 / P	FT / F	F
Shallow (S) 8 – 20 in. (20 – 50 cm.)	C_1	C_2 / P	C_3 / P	P	F	F
Very shallow (VS) < 8 in. (< 20 cm.)	C_1 / P	P	P	P	F	F

Fig. 4.5 Chart for determining land capability class according to the treatment-oriented scheme of Sheng (1972a; Table 4.3).

simple indices such as gully density. No attempt was made to map the factors that influence erosion and which could therefore provide the basis for predicting change. The approach to analysing change was the sequential survey in which mapping was carried out at regular intervals, using photography of different dates. In this way changes in gully density could be examined in relation to changing agricultural practices and increasing population pressure (Keech, 1969). More recently, the need for a dynamic approach in which both the erosion features and the factors influencing them are mapped and relationships sought has become apparent.

Because of the similarity of content in the studies of soil erosion and fluvial geomorphology, it seems logical to seek a method for a dynamic approach to erosion survey in the various systems of geomorphological mapping. A geomorphological map shows the form of the land surface, the properties of the soils and rocks beneath the surface, and the kind and magnitude of the geomorphological processes at work. Not surprisingly, most mapping systems are too complex for applied use and special-purpose geomorphological maps have to be produced for specific tasks (Demek, 1971). In the ITC System of Geomorphological Survey (Verstappen and van Zuidam, 1968) three maps are produced: a general geomorphological map, a morpho-conservation map, and a hydrological map. The morpho-conservation map is intended for erosion evaluation and has been so used in southern Italy (van Genderen, 1970; Rao, 1975). On the map are shown slope steepness, slope shape, present landuse, areas of rill and gully erosion and areas of mass movement.

A geomorphological mapping system for soil erosion survey has been derived from the ITC System and the work of Gerlach and Niemirowski (1968) by Williams and Morgan (1976). The aim is to portray information on the distribution and type of erosion, erosivity, runoff, slope length, slope steepness, slope curvature in profile and plan, relief, soil type and landuse. As much detail as possible is shown on a single map but to avoid clutter overlays can be used and are recommended for erosivity, soils and slope steepness. The legend (Fig. 4.6) is designed for use with colour: blue for water

SYMBOL	FEATURE	COLOUR
~~	Perennial water course	blue
~-~-~	Seasonal water course	blue
---	Crest line	brown
⌣	Contour line	brown
vvvvvv	Major escarpment	brown
ʊʊʊʊʊ	Convex slope break	brown
xvxvx	Concave slope break	brown
—#—#—	Waterfall	blue
—+—	Rapids	blue
— — —	Edge of flood plain	blue
▼▼▼▼	Edge of river terrace	blue
—·—·—	Back of river terrace	blue
ⱶ ⱶ ⱶ	Swamp or marsh	blue
⌶⌶⌶⌶⊃	Active gully	red
⌶⌶⌶⌶⊃	Stable gully	blue
↘ ↘ ↘ ↘	Active rills	red
≡ ≡ ≡	Sheetwash/rainsplash (inter-rill erosion)	red
—↙	River bank erosion	red .
⌒ᴖᴖᴖᴖᴖ	Landslide or slump scar	red
⊂ᴄᴄ	Landslide or slump tongue	red
ᴄᴄᴄᴄᴄ	Small slides, slips	red
⟝⟝⟝	Colluvial or alluvial fans	brown
▓▓▓▓▓	Sedimentation	brown
—·—·—	Landuse boundary (landuse denoted by letter e.g. R – rubber; F – forest; P – grazing land; L – arable land.)	green
———	Roads and tracks	black
—+—+—	Railway	black
⊥ ⊥ ⊥	Cutting	black
▼▼▼▼	Embankment	black
▪·▪·▪	Buildings	black
⌒⌐	Terrace	black
═══	Waterway	black

SLOPES

☐	0°–1°
▒	2°–3°
▨	4°–8°
▓	9°–14°
■	15°–19°
▦	over 19°

Fig. 4.6 Legend for mapping soil erosion.

features, including river terraces; brown for relief features such as contours and crest lines; red for accelerated erosion by water or wind; yellow for aeolian features; green for plant covers; and black for man-made features, including badly located and ill designed conservation structures.

Although much information is obtained from aerial photographs, this needs to be checked and supplemented by additional data collected in the field. The severity of erosion can be rated by a simple scoring system taking account of the exposure of tree roots, crusting of the soil surface, formation of splash pedestals, the size of rills and gullies, and the type and structure of the plant cover (Table 4.4). Observations are made using quadrat sampling over areas of 1 m², for ground cover, crusting and depth of ground lowering, 10 m² for shrub cover, and 100 m² for tree cover and the density of rills and gullies.

Table 4.4   Coding system for soil erosion appraisal in the field

CODE	INDICATORS
0	No exposure of tree roots; no surface crusting; no splash pedestals; over 70% plant cover (ground and canopy).
½	Slight exposure of tree roots; slight crusting of surface; no splash pedestals; soil level slightly higher on upslope or windward sides of plants and boulders; 30–70% plant cover.
1	Exposure of tree roots, formation of splash pedestals, soil mounds protected by vegetation, all to depths of 1–10mm; slight surface crusting; 30–70% plant cover.
2	Tree root exposure, splash pedestals and soil mounds to depths of 1–5cm; crusting of surface; 30–70% plant cover.
3	Tree root exposure, splash pedestals and soil mounds to depths of 5–10cm; 2–5mm thickness of surface crust; grass muddied by wash and turned downslope; splays of coarse material due to wash or wind; less than 30% plant cover.
4	Tree root exposure, splash pedestals and soil mounds to depths of 5–10cm; splays of coarse material; rills up to 8cm deep; bare soil.
5	Gullies; rills over 8cm deep; blow-outs and dunes; bare soil.

In interpreting the results of field survey, it is important to place the data in its time perspective. This is clearly indicated by the results of a survey of rangelands in northern Mexico (Table 4.5). In the study area the land divides into four main units: high, rocky mountains; dissected lower sierra; flatter piedmont slopes or *mediano*; and the alkali flats or *bajío*. Grazing is on a rotational basis, utilizing the bajío in the summer, the sierra in the winter, and the mediano in the autumn and spring. At the time of survey in late August 1976, a surprisingly poor correlation was obtained between erosion severity and the proportion of bare ground. But, on the sierra and mediano pastures, a good correlation was found between erosion severity and the percentage of the ground covered by grass and trees. The bajío, however, had a better grass cover than would be expected from its erosion coding. This reflects not only the erodibility of the bajío but also that it had remained ungrazed for the previous nine months, giving time for the grass to recover. The most severe erosion was observed on land recently grazed, with poor grass cover and no trees. That this is a direct result of greater runoff following a reduction in the vegetation cover, exposing the surface to livestock trampling and rain beat, is indicated by the greater compaction and crusting found in these areas which

Table 4.5   Field survey of rangelands in northern Mexico

REGION		% PLANT COVER				Grass+	% Bare	Erosion
	Transect	Grass	Forbs	Shrubs	Trees	Trees	ground	code
Sierra	1	22	4	8	13	35	65	0.31
	2	18	3	6	11	29	74	0.75
	3	25	2	14	1	26	63	1.00
	4	18	4	9	0	18	69	4.12
Mediano	5	28	3	5	0	28	70	1.33
	6	19	5	6	0	19	67	1.83
Bajío	7	48	0	0	0	48	52	1.66
	8	53	0	0	0	53	47	0.66

% Plant cover and % bare ground do not always total 100 because the same area of land may be covered by grass and a tree canopy.

% Bare ground includes rocky and stony surfaces and bare soil.

Erosion code is based on Table 4.4.

Data are averages of 8 sample sites on each transect on the sierra, 6 on the mediano and 3 on the bajío.

Transects are aligned across the slope on the contour and arranged from high (1) to low (8) ground.

Field survey was carried out in late August 1976 by the author, Sr L. Carlos Fierro and Sr J. Jabalera Ramos.

causes a lowering of the infiltration rate. With a change in rangeland condition from excellent pasture to bare soil, infiltrometer tests show that the amount of water which can infiltrate the soil from a dry state is reduced over 60 min from 280 mm to 110 mm on the sierra and from 360 mm to 105 mm on the mediano, and over 150 min from 267 mm to 130 mm on the mediano and from 150 mm to 73 mm on the bajío (Josué Martínez and González, 1971; Sánchez Muñoz and Valdés Reyna, 1975).

The maps produced in soil erosion surveys can be used in several ways. First, they depict the location and nature of erosion and distinguish between natural and accelerated erosion. Spatial variations in erosion intensity can be examined relative to the topographical, ecological and cultural information shown on the maps. Second, they not only indicate the location of an erosion problem but also enable the significance of that location in an erosion sequence to be appreciated. Often, complete sequences from erosion to deposition can be discerned in downslope, downstream or downwind directions. An understanding of such sequences is essential for predicting the broader, regional effects of changes in landuse or the installation of conservation works. Third, length, area and height measurements can be made from the maps. These, together with the knowledge of relief which the maps provide, are essential information for designing such conservation measures as terraces, grass waterways and contour bunds. Fourth, the maps contain much of the material required for assessing land capability. Thus, through erosion survey, investigations of the severity of erosion, land capability and conservation treatment can be better linked, aiding the soil conservationist in the delineation of areas suitable for arable farming, grazing and non-agricultural activities.

In addition to these four uses, the maps provide the foundation for estimating erosion risk and determining the chief factors influencing soil loss. To a limited extent these objectives can be achieved by employing simple indices such as gully density or a factorial scoring system similar to the one described earlier. More valuable, however, is to incorporate the data contained on the maps in models of the soil erosion system.

CHAPTER 5
MODELLING
SOIL EROSION

The techniques described in the last chapter enable the risk of soil erosion to be assessed. Before planning conservation work it is helpful if this assessment can be transformed into a statement of how fast soil is being eroded. Estimates of the rate of soil loss may then be compared with what is considered acceptable. In addition, it is useful if the effects on erosion rates of different conservation strategies can be determined. What

Table 5.1 Types of models

TYPE	DESCRIPTION
PHYSICAL	Scaled-down hardware models usually built in the laboratory; need to assume dynamic similitude between model and real world.
ANALOGUE	Use of mechanical or electrical systems analogous to system under investigation, e.g. flow of electricity used to simulate flow of water.
DIGITAL	Based on use of digital computers to process vast quantities of data.
(a) Deterministic	Based on mathematical equations to describe the processes involved in the model, taking account of the laws of conservation of mass and energy.
(b) Stochastic	Based on generating synthetic sequences of data from the statistical characteristics of existing sample data; useful for generating input sequences to deterministic and parametric models where data only available for short period of observation.
(c) Parametric	Based on identifying statistically significant relationships between assumed important variables where a reasonable data base exists. Three types of analysis are recognized: *black-box*: where only main inputs and outputs are studied; *grey-box*: where some detail of how the system works is known; *white-box*: where all details of how the system operates are known.

After Gregory and Walling (1973).

is required, therefore, is a method of predicting soil loss under a wide range of conditions.

Ideally, a predictive technique should satisfy the conflicting requirements of reliability, universal applicability, easy usage with a minimum of data, comprehensiveness in terms of the factors included, and the ability to take account of changes in landuse and conservation practice. Because of the complexity of the soil erosion system, with its numerous interacting factors, the most promising approach for developing a predictive procedure lies in formulating a conceptual model of the erosion process.

Most of the models used in soil erosion studies are of the parametric grey-box type (Table 5.1). They are based on defining the most important factors, measuring them, and, using statistical techniques, relating them to measurements of soil loss. In recent years it has been realized that this approach is less than satisfactory in meeting another important objective in formulating models, that of increasing our understanding of how the erosion system functions and responds to changes in the controlling factors. Greater emphasis is being placed at present on developing white-box parametric models and deterministic models. Along with this goes a switch from using statistical techniques to employing mathematical ones frequently requiring the solution of partial-differential equations.

Before selecting an approach to modelling, the scale of operation must be considered. The detailed requirements for modelling erosion over a large drainage basin differ from those demanded by models of soil loss from a short length of hillslope or at the point of impact of a single raindrop. Scale influences the number of factors which need to be incorporated in the model, which ones can be held constant and (Table 1.1) which can be designated primary factors around which the model must be constructed.

The success of all predictive techniques depends on how closely the estimated rates of erosion compare with those measured in the field. The development of models is thus necessarily tied up with data collection on rates of soil loss. Data are needed not only to test the performance of models but also, especially in the case of parametric models, to formulate them.

5.1. DATA COLLECTION

Data on soil erosion and its controlling factors can be collected in the field or, for simulated conditions, in the laboratory. Whether field or laboratory experiments are used depends on the objective. For realistic data on soil loss, field measurements are the most reliable, but because conditions vary in both time and space, it is often difficult to determine the chief causes of erosion or to understand the processes at work. The mechanics of erosion are best studied in the laboratory where the effects of many factors can be controlled. Because of the artificiality of laboratory experiments, however, some confirmation of their results in the field is desirable.

5.1.1. Field experiments

Field experiments may be classified into two groups: those carried out at permanent research or experimental stations and those designed to assess erosion at a number of sample sites over a large area.

Work at experimental stations is based on bounded runoff plots of known area, slope steepness, slope length and soil type from which both runoff and soil loss are monitored. The number of plots depends upon the pupose of the experiment but usually allows for at least two replications. Thus, to assess the rate of erosion under two crop types would require a minimum of four plots. The introduction of a soil variable would increase the number to eight, if two soil types were involved. The plot layout is designed to give a random pattern with respect to variables not directly related to the investigation (Fig. 5.1).

The standard plot is 22 m long and 1.8 m wide although other plot sizes are sometimes used. The plot edges are made of sheet metal, wood or any material which is

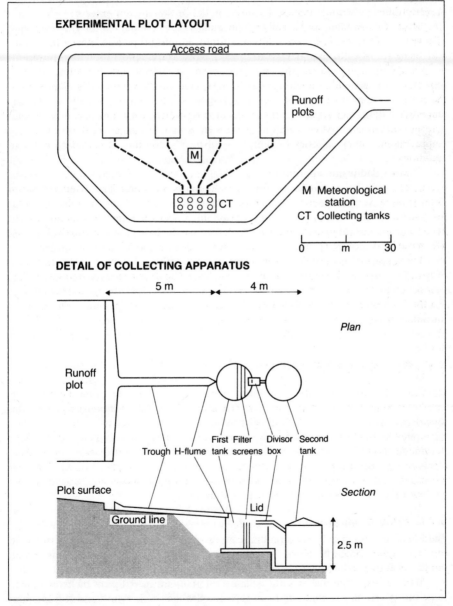

Fig. 5.1 Typical layout of runoff plots at a soil erosion and conservation research station (after Hudson, 1965).

stable, does not leak and is not liable to rust. The edges should extend 150–200 mm above the soil surface. At the downslope end is positioned a collecting trough or gutter, covered with a lid to prevent the direct entry of rainfall, from which sediment and runoff are channelled into collecting tanks. For large plots or where runoff volumes are very high, the overflow from a first collecting tank is passed through a divisor which splits the flow into equal parts and passes one part, as a sample, into a second collecting tank. Examples are the Geib multislot divisor which samples one-fifth of the overflow and the Coshocton wheel which samples one-hundredth. On some plots, prior to passing into the first collecting tank, the runoff is channelled through a flume, normally an H-flume, where the discharge is automatically monitored. Rainfall is measured with both standard and autographic gauges adjacent to the plots.

Although the bounded runoff plot gives probably the most reliable data on soil loss per unit area, there are several sources of error involved with its use (Hudson, 1957). These include silting of the collecting trough and pipes leading to the tanks, inadequate covering of the troughs against rainfall, and the maintenance of a constant level between the soil surface and the sill or lip of the trough. Other problems are that runoff may collect along the boundaries of the plot and form rills which would not otherwise develop, and that the plot itself is a partially closed system, being cut off from the input of sediment and water from upslope.

An alternative method of measuring sediment loss and runoff has been developed by Gerlach (1966) using simple metal gutters, 0.5 mm long and 0.1 m broad, closed at the sides and fitted with a movable lid. An outlet pipe runs from the base of the gutter to a collecting bottle. In a typical layout, two or three gutters are placed side-by-side across the slope and groups of gutters are installed at different slope lengths, arranged *en echelon* in plan to ensure a clear run to each gutter from the slope crest (Gerlach, 1967; Morgan, 1977). Because no plot boundaries are used, edge effects are avoided. It is normal to express soil loss per unit width but if an areal assessment is required, it is necessary to assume a catchment area equal to the width of the gutter times the length of the slope. A further assumption is that any loss of water and sediment from this area during its passage downslope is balanced by inputs from adjacent areas. This assumption is reasonable if the slope is straight in plan. On slopes curved in plan, the catchment area must be accurately surveyed in the field. This disadvantage is offset by the flexibility of monitoring soil loss at different slope lengths and steepnesses which the method provides. The gutters have been used, for example, to study soil loss along crest lines where the latter fall in altitude on spur-ends (Yaïr, 1972). The method also allows the measurement of erosion within an open system.

Because of their cheapness and simplicity, Gerlach troughs can be employed for sample measurements of soil loss at a large number of selected sites over a large area. They are thus suitable for erosion studies on a watershed scale. Small bounded plots of 4 m² and 8 m² have been similarly used by Soons and Rainer (1968) in New Zealand and Kellman (1969) in the Philippines respectively.

Investigations of sediment production must be carried out on hillslopes and in rivers if a full picture of erosion over a large watershed is to be obtained. In many cases, the amount of sediment leaving a drainage basin is less than that removed from the steeper sections of the hillsides. This points to deposition of material at the base of the hillslopes and on flood plains. Where more sediment leaves a basin than is contributed to the rivers from the hillslopes, the balance is derived from erosion of the river banks.

The selection of measurement sites to establish the pattern of sediment movement poses a problem of sampling. One approach is to divide a large watershed into subbasins and set up recording stations on the rivers at the mouth of each. Discharge is measured automatically using weirs and water depth recorders. Suspended sediment concentrations are determined from water samples taken at set times with buckets or specially designed sediment samplers, or they are monitored continuously by recording the turbidity of the water (Gregory and Walling, 1973). Within each sub-basin, slope profiles are chosen at regular intervals or at random along the mid-slope line which lies half-way between the main river and the divide. Gerlach troughs are installed either randomly or in set positions on the profiles. A typical layout is shown in Fig. 5.2.

The techniques for measuring wind erosion are less well established than those for monitoring water erosion. Various types of traps are used to catch sand moving in a band of unit width. Horizontal traps consist of troughs set in the ground level with the surface and parallel with the direction of the wind. The trap is sometimes divided into compartments so that rolling and saltating sand particles fall into different compartments according to their length of hop. Alternatively, several traps of different lengths may be used. The horizontal traps have the advantage of minimum interference with the wind but a considerable length is required to collect a representative sample. Studies by Borsy (1972) in maize fields on sand dunes near Hajdúböszörmény in the Duna-Tisza interfluve region of Hungary show that in a storm of 2 h 30 min, with a wind velocity of 8–10 m s^{-1}, the amount of sand collected in 10 cm wide horizontal traps varied from 218 g for 1 cm length, to 370 g for 50 cm length and 570 g for 1 m length. Vertical traps consist of a series of boxes placed one above the other so as to catch all the particles moving at different heights. These traps are unsatisfactory aerodynamically, however, because it is difficult to supply a sufficiently large exhaust to permit the free flow of air. The build up of back pressure causes resistance to the wind which is deflected from the traps. A problem with both horizontal and vertical traps (Fig. 5.3) is that they cannot be easily reoriented as the wind direction changes. In order to collect sand from all directions, De Ploey (1977), at the Kalmthout Dune Station, near Antwerp, uses a circular collector consisting of a 60 cm high pile of 16 cm diameter cake tins.

5.1.2. Laboratory experiments

Many laboratory studies centre around the use of a rainfall simulator which is designed to produce a storm of known energy and drop-size characteristics which can be repeated on demand. Many simulators are available but none accurately reproduce all the properties of natural rain (Hall, 1970). There is insufficient height in a laboratory for water drops to achieve terminal velocity during fall so their kinetic energy is low. To overcome this, water is released from low heights under pressure. This results in too high an intensity and, because the increase in pressure produces small drop sizes, unrealistic drop-size distributions. The intensity can be brought down by reducing the frequency of rain striking the target area, either by oscillating the spray over the target or by intermittently shielding the target from the spray. Rainfall simulators are classified according to the drop-formers used. These range from hanging yarn (Ellison, 1947) to tubing tips, either hypodermic needles or capillary tubes (De Ploey, Savat and Moeyersons, 1976), and nozzles (Morin, Goldberg and Seginer, 1967).

Where the target is in the form of a small soil plot, the rainfall simulator may be supplemented by a device to supply a known quantity of runoff at the top of the plot, instead of relying solely on runoff resulting from the rainfall. This facility is helpful for studies of the hydraulics of overland flow during rain (Savat, 1977).

Almost all the basic studies on wind erosion have been carried out in the laboratory in wind tunnels. Wind is supplied by a fan which either sucks or blows air through the

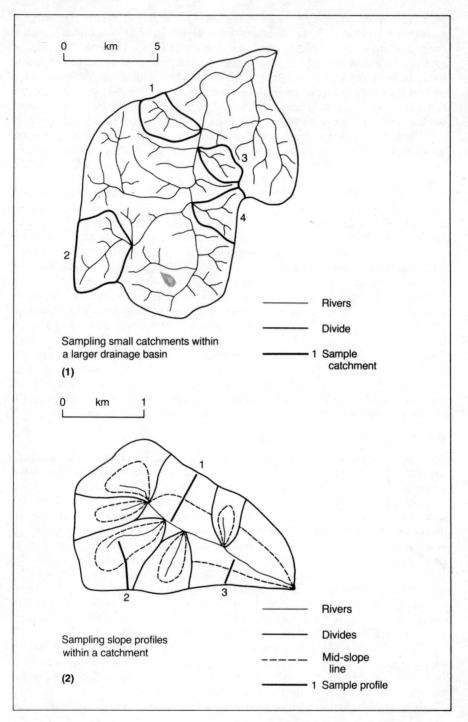

0   km   5

Sampling small catchments within
a larger drainage basin

(1)

— Rivers

— Divide

■ 1 Sample
      catchment

0   km   1

Sampling slope profiles
within a catchment

(2)

— Rivers

— Divides

--- Mid-slope
      line

■ 1 Sample profile

Fig. 5.2 Sampling scheme and equipment for measuring soil loss from hillslopes.

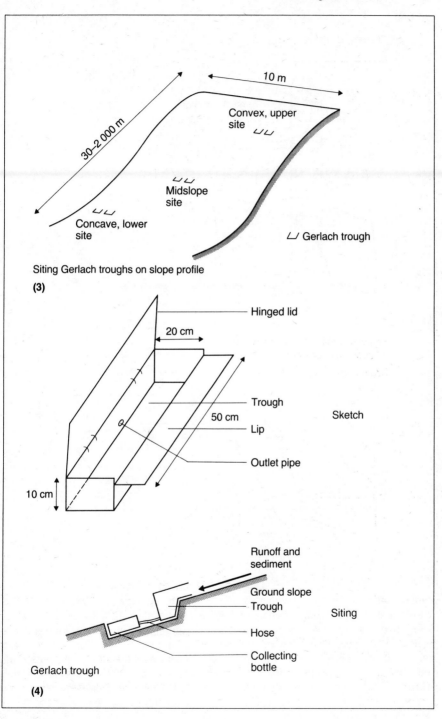

10 m

Convex, upper
site   ⌐⌐

30–2 000 m

Midslope
site

Concave, lower
site

⌐ Gerlach trough

Siting Gerlach troughs on slope profile

(3)

Hinged lid

20 cm

Trough

50 cm

Lip

Outlet pipe

Sketch

10 cm

Runoff and
sediment

Ground slope
Trough

Hose

Collecting
bottle

Siting

Gerlach trough

(4)

Fig. 5.2 *(cont.)*

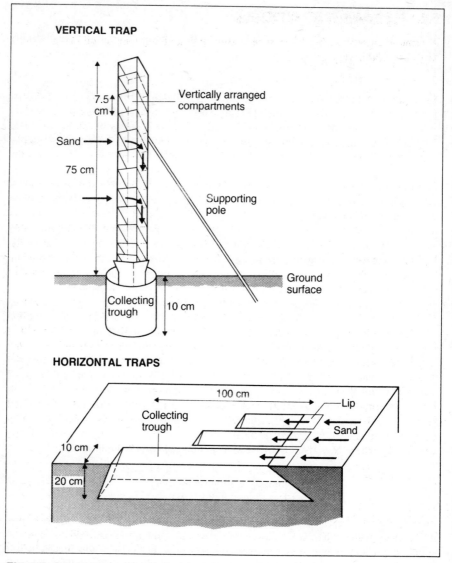

Fig. 5.3 Sand traps.

tunnel. The tunnel is shaped so that air enters through a convergence zone, where flow is constricted, passes through the test section, and leaves through a divergence zone, in which flow is diffused. A honeycomb shield at the outlet traps most of the sand particles whilst allowing the air to blow through without a build-up of back pressure. Realistic wind profiles are produced only in the test section and the value of the tunnel depends on the length of this. Small tunnels, such as the one described by De Ploey and Gabriels (in press) with a test section only 1.5 m long, do not permit a satisfactory sand flow to be attained. At least a 15 m long section is required for this though Bagnold (1937) was able to simulate sand flow in a 10 m long tunnel by feeding a stream of sand into its mouth.

5.2. PARAMETRIC MODELS

The simplest model is a black-box type relating sediment loss to either rainfall or runoff. A typical relationship is:

$$Qs = aQw^b, \tag{5.1}$$

where Qs is sediment discharge and Qw is water discharge. Jovanović and Vukčević (1958), using data for sixteen gauging stations in Yugoslavia, established that $b = 2.25$ whilst, according to Leopold, Wolman and Miller (1964), b ranges in value from 2.0 to 3.0. The value of a is an index of erosion severity. Thus, in the Yugoslav example, $a > 0.000\ 7$ denotes excessive soil loss and $a < 0.000\ 3$ indicates a low erosion rate. Parsons, Apmann and Decker (1964) noted a fall in the value of a from 1.46 to 0.86 for the Buffalo River, New York, following channel stabilization. Since the latter study is based on sediment concentration (ppm) rather than sediment discharge (kg s⁻¹), these values should be increased by 1.0 to compare with the Yugoslav work. It is not always possible to establish the values with confidence. The relationship between sediment and water discharge may vary with the volume of runoff and therefore change seasonally. During a single storm the value of b often differs on the rise of a flood wave from that on the recession (Gregory and Walling, 1973). The main disadvantage of this type of model, however, is that it gives no indication of why erosion takes place.

Greater understanding of the causes of erosion is achieved by grey-box models. These are usually empirical and, as seen in Chapter 2, frequently culminate in expressing the relationship between sediment loss and a large number of variables with a regression equation. One example is equation (4.1) of Fournier (1960) relating mean annual sediment yield to rainfall, altitude and slope. Another is the following equation derived by Walling (1974) which explains 96 per cent of the variation in suspended sediment yields (kg) of single storm runoff events for a small catchment near Exeter in terms of storm duration (DUR; h), peak runoff (Qw; l s⁻¹), peak surface runoff, defined as runoff less base flow (Qq; l s⁻¹), total surface runoff (QQ; mm), flow level preceding the hydrograph rise (Qap; l s⁻¹), and the day of the year (DY) evaluated as sine (radians) $2\pi D/365$, where D is the day of the year numbered from 1 January:

$$\begin{aligned}\log Qs = &- 1.140\ 2 - 0.052\ 4DUR - 0.776\ 4\ \log_{10}Qw \\ &+ 1.373\ 5\ \log_{10}Qq + 0.989\ 2\ \log_{10}QQ \\ &- 0.496\ 1\ \log_{10}Qap + 0.269\ 3\ DY.\end{aligned} \tag{5.2}$$

Many of the variables included in this type of model are themselves intercorrelated and it is often difficult to identify the most important ones. Thus, although, as in this case, the equation may have a highly explanatory and, therefore, predictive value in a statistical sense, it has a low explanatory value conceptually. Walling (1974) suggests using a principal components analysis to reduce the number of variables by identifying the redundant ones. This technique was employed successfully by Douglas (1968a) in a study of suspended sediment yields of rivers in northern Queensland. Ten possible controlling factors were reduced to four basic factors: wetness, drainage basin morphology, lithology and terrain roughness. He derived the following equation to explain 82 per cent of the variation in annual suspended sediment yield (SS; m³ km⁻²):

$$\begin{aligned}\log SS = &- 8.73 + 3.81\ \log_{10}QwA - 1.54\ \log_{10}R/L \\ &+ 4.82\ \log_{10}DD,\end{aligned} \tag{5.3}$$

where QwA is mean annual runoff (mm), R/L is the relief/length ratio of the drainage basin (ft mi⁻¹) and DD is drainage denisty (ft mi⁻²).

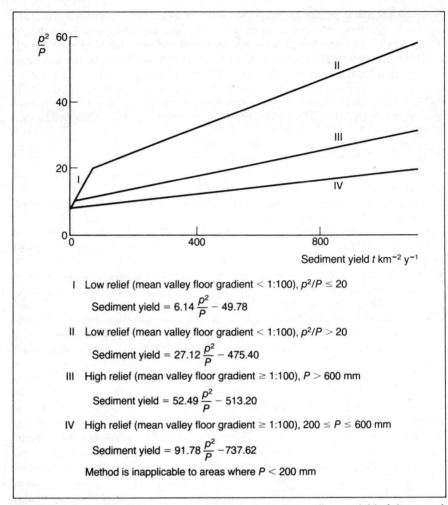

Fig. 5.4 Relationship between mean annual suspended sediment yield of rivers and rainfall aggressiveness (after Fournier, 1960).

A further problem of these empirical models is that the equations cannot be extrapolated beyond the data range with confidence, either to more extreme events or to other areas. Although equation (4.1) is often assumed to have general validity, its limitations were recognized by Fournier (1960) who also produced four other regression equations relating sediment yield to the p^2/P index for specific conditions of relief and climate (Fig. 5.4). Probably the nearest approach to an equation with world-wide applicability is that developed by Douglas (1967b) relating mean annual suspended sediment yield (m³ km⁻²) to effective precipitation (*PE*):

$$Qs = \frac{1.631(0.039\ 37PE)^{2.3}}{1 + 0.007(0.039\ 37PE)^{3.3}} \tag{5.4}$$

The numerator in this equation represents the direct erosive effect of rainfall whilst the denominator attempts to take account of the protective effect of the plant cover.

All the predictive equations discussed so far refer to drainage basins and do not provide a suitable technique for assessing soil loss from smaller areas such as hillslopes and fields. The first attempt to develop a soil loss equation for hillslopes was that of Zingg (1940) who related erosion to slope steepness and slope length (equation 3.8). Further developments led to the addition of a climatic factor based on the maximum 30-min rainfall total with a two-year return period (Musgrave, 1947), a crop factor, to take account of the protection-effectiveness of different crops (Smith, 1958), a conservation factor and a soil erodibility factor. Changing the climatic factor to the rainfall erosivity index (R) ultimately yielded the Universal Soil-Loss Equation (Wischmeier and Smith, 1962):

$$E = R.K.L.S.C.P., \tag{5.5}$$

where E is mean annual soil loss (t ac^{-1} y^{-1}). The derivation of the factors in the equation is as follows.

R. This is the rainfall erosivity index which is equal to the mean annual erosivity value (section 3.1.2) divided by 100:

$$R = \frac{EI_{30}}{100} \tag{5.6}$$

K. This is the soil erodibility index (section 3.2) defined as mean annual soil loss per unit of erosivity for a standard condition of bare soil, no conservation practice, 5° slope of 22 m length.

LS. The factors of slope length (L) and slope steepness (S) are combined in a single index. A value of 1.0 applies to the standard 5° slope, 22 m long. The appropriate value can be obtained from nomographs (Hudson, 1971) or from the equation:

$$LS = \sqrt{\frac{L}{100}} \; (0.136 + 0.097 \, S + 0.013 \, 9 \, S^2), \tag{5.7}$$

where L is in m and S in per cent.

C. This is the crop factor. It represents the ratio of soil loss under a given crop to that from bare soil. Since soil loss varies with erosivity and the morphology of the plant cover (section 3.4), it is necessary to take account of changes in these during the year in arriving at an annual value. The year is divided into periods corresponding to different stages of crop growth. For annuals, these are: (i) fallow – turn-ploughing to seeding; (ii) seedbed – seeding to one month thereafter; (iii) establishment – 1 to 2 months after seeding; (iv) growing period – from establishment to harvest; and (v) residue or stubble. For the last period, three options are considered: leaving the residue in the field without seeding an off-season crop; leaving the residue and seeding an off-season crop; and removing the residue and leaving the ground bare. For a given crop, separate ratio values are obtained for each period from tables (Wischmeier and Smith, 1965) summarizing data collected over many years by the United States Soil Conservation Service at their experimental stations. The values vary not only with the crop but also, for a single crop, with yield, plant density and the nature of the previous crop. The individual values for each period are weighted according to the percentage of the mean annual erosivity falling in that period and summed to give the annual value for C.

P. This is the conservation practice factor. Values are obtained from tables of the ratio of soil loss where contouring and contour strip-cropping are practised to that where they are not. With no conservation measures, the value of P is 1.0. Where terracing is adopted, the value for strip-cropping is used for the P factor and the LS index is adjusted for the slope length which represents the horizontal spacing between the terraces.

Table 5.2 Prediction of soil loss using the Universal Soil-loss Equation

Calculation of mean annual soil loss on a 100m long slope of $7°$ on soils of the Rengam Series under maize cultivation with contour bunds, near Kuala Lumpur.

BASIC EQUATION

Soil loss $= R \times K \times LS \times C \times P$

ESTIMATING R (rainfall erosion index).

Method 1:	Mean annual precipitation	$= 2\,695$mm
	From Roose (1975), mean annual erosivity (EI_{30})	$= 2\,695 \times 50$
		$= 134\,750$
	Rainfall erosion index	$= \dfrac{134\,750}{100}$
		$= 1\,347.5$
Method 2:	From Morgan (1974), mean annual erosivity (KE >25)	$= 9.28$ (mean annual precipitation) — $8\,838$
		$= 9.28(2\,695) - 8\,838$
		$= 16\,171.6$ J m^{-2}
	Convert to N m	$= 16\,171.6$ N m m^{-2}
	Convert to kgf m	$= 16\,171.6 \times 0.102$
		$= 1\,649.5$
	Multiply by mean annual I_{30} (taken as roughly equal to daily rainfall total with 2-y return period, which for Kuala Lumpur is 85mm).	$= 1\,649.5 \times 85$
		$= 140\,207.5$
	Rainfall erosion index	$= \dfrac{140\,207.5}{100}$
		$= 1\,402.1$
Best estimate (average of two methods)		$= \underline{1\,370}$

ESTIMATING K (soil erodibility index).

From Whitmore and Burnham (1969), the soils have 43% clay, 8% silt, 9% fine sand and 40% coarse sand; organic content about 3%.

Using the nomograph (Fig. 3.2), gives a first approximation K value $= \underline{0.05}$

ESTIMATING LS (slope factor)

For slope length (L) and slope steepness (S) in m and per cent respectively,

$$LS = \sqrt{\frac{L}{100}}\,(0.136 + 0.097S + 0.013\,9S^2)$$

For $L = 100$ m and $S = 12\%$ (approximation of $7°$)

$$LS = \sqrt{\frac{100}{100}}\,(0.136 + 0.097 \times 12 + 0.013\,9 \times 12^2)$$

$$= 1.0 \times 3.302$$

$$LS = \underline{3.3}$$

ESTIMATING C (crop practice factor)

According to Roose (1975), the C value for maize ranges between 0.4 and 0.9, depending upon the cover.

continued

Table 5.2 (*cont.*)

During the three month period from seeding to harvest, the cover varies from 9 to 45 per cent in first month, to 55 to 93 per cent in second month, and 45 to 57 per cent in third month.

Therefore, assume C values of 0.9, 0.4 and 0.7 for the three respective months.

Maize can be planted at any time of the year in Malaysia but assume planting after the April rains, allowing growth, ripening and harvesting in June and July which are the driest months. Land is under dense secondary growth prior to planting ($C = 0.001$) and allowed to revert to the same after harvest (assume $C = 0.1$).

Of the mean annual precipitation, 32 per cent falls between January and April inclusive, 10 per cent in May, 6 per cent in June, 7 per cent in July, and 45 per cent between August and December. Assuming that erosivity is directly related to precipitation amount, these values can be used to describe the distribution of erosivity during the year.

From these data, the following table is constructed:

MONTHS	C VALUE	ADJUSTMENT FACTOR (% R VALUE)	WEIGHTED C VALUE (col 2×col 3)
January–April	0.001	0.32	0.000 32
May	0.9	0.10	0.09
June	0.4	0.06	0.024
July	0.7	0.07	0.049
August–December	0.1	0.45	0.045
		Total	0.208 32

C factor for the year = 0.208

ESTIMATING P (conservation practice factor)

From Roose (1975), P value for contour bunds = 0.3

Soil loss estimation

Mean annual soil loss = $1370 \times 0.05 \times 3.3 \times 0.208 \times 0.3$

(t ac^{-1} y^{-1})   = 14.10

As the example in Table 5.2 shows, the equation is normally used to predict soil loss. Since it was derived from and tested on data from experimental stations in the United States which, when combined, represent over 10 000 years of records, it is widely accepted as reliable. It has become the standard technique of soil conservation workers (F.A.O., 1965; Hudson, 1971). The equation can be rearranged so that, if an acceptable value of E is chosen, the slope length (L) required to reduce soil loss to that value can be calculated. In this way, appropriate terrace spacings can be determined. Alternatively, the C value may be predicted and the tables searched to find the most suitable cropping practice which will give that value.

Although the equation is described as universal, its data base, though extensive, is restricted to the United States east of the Rocky Mountains. The base is further restricted to slopes where cultivation is permissible, normally 0 to 7°, and to soils with a low content of montmorillonite. Several attempts have been made to apply the equation more widely. Hudson (1961) has modified it for Rhodesia and Roose (1975) has investigated its applicability to the Ivory Coast and, in so doing, has concentrated on defining values for the C factor for crops which are not grown in the United States.

In addition to these practical limitations, there are theoretical problems with the equation. There is considerable interdependence between the variables and some are counted twice. For instance, rainfall influences the R and C factors and terracing the L and P factors. Other interactions between factors, such as the greater significance of slope steepness in areas of intense rainfall (section 3.3), are ignored. The rainfall erosion index is based on studies of drop-size distributions of rains which, as seen in section 3.1.2, may have limited applicability. One important factor to which soil loss is closely related, namely runoff, is omitted. To overcome this, Foster, Meyer and Onstad (1973) have suggested replacing the R factor with an energy term, W, which is a function of rainfall and runoff energy, defined as:

$$W = 0.5R + 15Qq_P^{1/3}, \tag{5.8}$$

where R is the rainfall erosivity factor, Q is storm runoff (in) and q_P is storm peak runoff rate (in h^{-1}).

A similar soil loss equation has been developed for wind erosion (Chepil and Woodruff, 1963) taking account of soil erodibility, wind energy, surface roughness, length of open wind blow and the vegetation cover. The equation requires numerous tables and nomographs to operate because the factors cannot be combined by multiplication as with the equation for water erosion. Because of this, it has limited practical value.

The closest approach to a white-box model for sediment production is the Stanford Sediment Model (Negev, 1967) which is an addition to the Stanford Watershed Model used to predict runoff. The model is best illustrated as a flow chart (Fig. 5.5). Rainfall, overland flow and channel flow comprise the inputs to the model and outputs consist of sediment removed from the hillslopes by overland flow and sediment derived from rill,

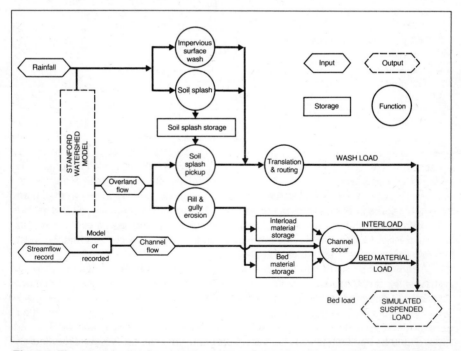

Fig. 5.5 Flow chart for the Stanford Sediment Model (after Gregory and Walling, 1973).

gully and channel erosion. The operation of the model is based on several functions, each of which describes a process in the erosion system and is expressed by an equation. Although the model applies to a drainage basin, separate models of the section covering erosion by overland flow are being developed (David and Beer, 1975).

5.3. DETERMINISTIC MODELS

Deterministic models are based on the laws of conservation of mass and energy. Most of them use a particular differential equation known as the continuity equation which is a statement of the conservation of matter as it moves through space over time. The equation can be applied to soil erosion on a small segment of a hillslope as follows. There is an input of material to the segment as a result of detachment of soil particles on the segment itself and an influx of sediment from the slope above. There is an output of

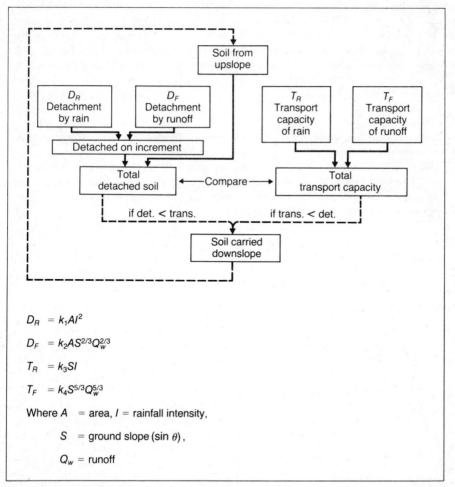

$D_R = k_1 A I^2$

$D_F = k_2 A S^{2/3} Q_w^{2/3}$

$T_R = k_3 S I$

$T_F = k_4 S^{5/3} Q_w^{5/3}$

Where A = area, I = rainfall intensity,

S = ground slope (sin θ),

Q_w = runoff

Fig. 5.6 Flow chart for the model of the processes of soil erosion by water (after Meyer and Wischmeier, 1969).

material through the processes of transport by rainsplash and runoff. If the transporting processes have the capacity to remove all the material supplied, there is a net loss of soil to the segment. If the transport capacity is insufficient, there is a net gain of soil. Thus, for the slope segment:

$$\text{input} - \text{output} = \text{loss or gain of matter}. \tag{5.9}$$

This approach, illustrated by a flow chart (Fig. 5.6), has been employed by Meyer and Wischmeier (1969), in a mathematical model designed to simulate soil erosion as a dynamic process. The operation of the model uses four equations to describe the separate processes of soil detachment by rainfall, soil detachment by runoff, transport capacity of rainfall and transport capacity of runoff. Using these equations and applying the model to consecutive downslope segments in turn, sediment can be routed and the pattern of erosion evaluated along a complete slope profile.

Because of its simplicity, the model has significant limitations. Steady-state conditions are assumed for rainfall intensity, infiltration rate and runoff rate. No other erosion processes and no weathering are allowed for. The soil is assumed bare of plant cover and no account is taken of tillage practices. Surface depression storage is ignored. There is no removal of material from the base of the slope and the altitudes of the highest and lowest points of the slope remain fixed through time. All these limitations are, of course, recognized by the authors. Improvements are being made to the model to turn it into a useful design tool (Foster and Meyer, 1975). Similar slope erosion models are being developed for a range of spatial and temporal scales (Kirkby, 1971; Ahnert, 1976).

Deterministic modelling must be seen as the ultimate objective of research into soil erosion models. At present, the work is in its infancy. Many of the models are simplistic, few have been tested against observed data and none have immediate practical value. When perfected, however, such models will provide a more comprehensive and widely applicable technique than parametric models for assessing the relative importance of the various factors of the soil erosion system. They will strengthen the basis on which strategies for conserving soil are designed.

CHAPTER 6
STRATEGIES FOR
EROSION CONTROL

The aim of soil conservation is to obtain the maximum sustained level of production from a given area of land whilst maintaining soil loss below a threshold level which, theoretically, permits the natural rate of soil formation to keep pace with the rate of soil erosion. From the discussion (Ch. 2) of the mechanics of the detachment and transport of soil particles by rainsplash, runoff and wind, it follows that the strategies for soil conservation must be based on covering the soil to protect it from raindrop impact; increasing the infiltration capacity of the soil to reduce runoff; improving the aggregate stability of the soil; and increasing surface roughness to reduce the velocity of runoff and wind. In this chapter the purpose and use of various conservation techniques are described under the widely-accepted headings of agronomic measures, soil manage-ment and mechanical methods. The detailed design procedures are, however, beyond the scope of this book and are well-covered in texts on soil conservation (Stallings, 1957; Schwab, Frevert, Edminster and Barnes, 1966; Hudson, 1971).

6.1. AGRONOMIC MEASURES

These are based on the role of the plant cover in reducing erosion (section 3.4). Because of differences in their density and morphology, plants differ in their effectiveness in protecting the soil from erosion. This can be shown by studying erosion rates measured from small plots in the field under different plant covers (F.A.O., 1965; Kellman, 1969; Roose, 1971; Temple, 1972a). Generally, row crops are the least effective and give rise to the more serious erosion problems. This is because of the high percentage of bare ground, particularly in the early stages of crop growth, and the need to prepare a seed bed. In designing a conservation strategy based on agronomic measures, row crops must be combined with protection-effective crops in a logical cropping pattern.

6.1.1. Rotation

The simplest way to combine different crops is to grow them consecutively in rotation. The frequency with which row crops are grown depends upon the severity of erosion. Where erosion rates are low, they may be grown every other year, but in very erodible areas, they may be permissible only once in five or even seven years. A high rate of soil loss under the row crop is acceptable because it is counteracted by low rates under the other crops so that, averaged over a six or seven year period, the annual erosion rate remains low.

Suitable crops for use in rotations are legumes and grasses. These provide good ground cover, help to maintain or even improve the organic status of the soil, thereby contributing to soil fertility, and enable a more stable aggregate structure to develop in the soil. These effects are often sufficiently long-lasting as to reduce erosion and increase yield during the first year of row-crop cultivation but they rarely extend into the second year. For this reason, two continuous years of planting with a row crop should be avoided. Hudson (1971) shows that a rotation of tobacco–grass–grass–tobacco–grass–grass is more effective in Rhodesia than one of two consecutive years of tobacco followed by four years of grass. The respective mean annual soil loss rates are 1.2 and 1.5 kg m^{-2}. The same effect is illustrated more dramatically by Kellman (1969) for shifting cultivation in the Philippines where, during the cropping period, soil loss from land under rice averages 0.38 g m^{-2} day^{-1} on a new clearing but rises to 14.91 g m^{-2} day^{-1} on a clearing in its twelfth year of cultivation.

Rotation is commonly practised on grazing land, moving the stock from one pasture to another in turn, to give time for the grass to recover. Generally, grasslands should not be exploited to more than 40 to 50 per cent of their annual production (Fournier, 1972) and should be allowed to regenerate sufficiently to provide a 70 per cent ground cover at the times of erosion risk. Grazing land has to be very carefully managed. Whilst overgrazing can lead to deterioration of the rangeland and the onset of erosion, undergrazing can result in the loss of nutritious grasses, many of which regenerate more rapidly when grazed. Similarly, although erosion rates are high from burnt-over land, controlled burning is essential for the removal of undesirable plant species. Uncontrolled burning can prevent plants from re-establishing and, by increasing the extent of bare ground, result in serious erosion. Imeson (1971) attributes soil erosion on the North Yorkshire Moors to burning of the heather every two or three years. This frequency of burn compares with a six-year period necessary for *Calluna* to develop from seed to a height of 30 to 40 cm and provide a complete canopy to the ground surface. The greatest risk of erosion is on peaty soils where ground lowering rates up to 45 mm y^{-1} have been measured following a burn. Shallow, intermittent gullies, from a few decimetres to 4 m deep, cross the *Calluna* moorland. Where the vegetation cover is poor, these have been found to increase in width at rates of 17 to 80 mm y^{-1}. Clearly, burning must be practised on a rotational basis and only carried out with a frequency which will permit plants to regrow and maintain soil loss, when averaged over six or seven years, at an acceptably low level.

A critical factor in a rotational grazing system is the quality of the poorest rangeland. On the rangelands of northern Mexico, the recommended stocking rates to maintain the pasture in 'good' condition, a state defined locally as having at least 50 per cent desirable grasses in the vegetation cover, are one animal per 20 hectares in the mountains, one animal per 14 hectares on the mediano, and one animal per 14 hectares on the sierra and bajío (González, 1971). With the rotational pattern described in section 4.2.3 in operation, any upgrading of the mediano pastures will strain the capacity of the other pastures in the system. Similar strains are being experienced on the hill pastures in Wales where farmers, having improved the grasses of the lower slopes and valleys, have increased stocking densities, resulting in greater numbers of sheep being put to graze on the unimproved upland pastures during the summer. Overgrazing of these pastures may be one cause of soil erosion in the Clwyd Range.

The principles of rotation can be applied to other forms of landuse. The timber resources of forests are often exploited commercially by patch-cutting, also termed clear felling, on a rotational basis. Erosion rates are highest in the years immediately following logging operations but decline in subsequent years either with natural vegetation regrowth or replanting so that averaged over twelve years or more they may be little different from those of undisturbed land. Leaf (1970) shows that in Fool Creek, Colorado, in the Rocky Mountains, soil loss averaged 22.4 g m^{-2} y^{-1} whilst logging and road construction were in progress in the 1950s but, since 1958, a year after operations ceased, has averaged only 9.86 g m^{-2} y^{-1}. The latter figure compares with soil loss of 4.81 g m^{-2} y^{-1} from uncut forest land. Although, on this basis, patch-cutting may be an acceptable management technique, in very erodible areas, soil loss may be too severe during logging to permit its use. In mountainous areas of Peninsular Malaysia, slides and gullies develop rapidly when the forest is cleared (Berry, 1956; Burgess, 1971) and increased erosion and runoff may result in problems of sedimentation and flooding downstream. In these areas, selective felling, whereby only the mature trees are removed and the other trees remain to provide a plant cover, is better conservation practice. Rich (1972) has observed that selective felling does not significantly increase runoff compared with undisturbed forest areas. Since, in his study area of Arizona, annual precipitation averages less than 800 mm, he recommends managing the Ponderosa pine forests by clear cutting in order to increase water yield.

In recent years, as the problem of erosion in recreational areas has become recognized, interest has focused on whether rotational practices can be adapted to people. Although there is a reluctance to treat people in the same way as livestock, it appears to be increasingly necessary in heavily used areas to fence off sections of land and to close certain footpaths and access routes so that the vegetation cover has a chance to regenerate. In many upland areas, where grazing, forestry and recreation are complementary but sometimes competing activities, the prospect of applying a common conservation measure is extremely attractive.

6.1.2. Cover crops

Cover crops are grown as a conservation measure either during the off-season or as ground protection under trees. In the United States they are grown as winter annuals and, after harvest, are ploughed in to form a green manure. Typical crops used are rye, oats, hairy vetch, sweet clover and lucerne in the north, and Austrian winter peas, crimson clover, crotalaria and lespedeza in the south.

Ground covers are grown under tree crops to protect the soil from the impact of water drops falling from the canopy. They are particularly important with tall crops such as rubber where the height of fall is sufficient to cause the drops to approach their terminal velocity (section 3.4). The most common crops used in tropical areas are *Pueraria phaseoloides, Calopogonium mucunoides* and *Centrosema pubescens*. Although these covers have the advantages of rapid growth and retaining nutrients in the soil which would otherwise be removed by leaching, these are sometimes offset by problems. First, there is a risk that a satisfactory cover will not be attained. This is because, as the main crop becomes established, ground conditions change from strong, open sunshine to shade, causing some plants to die out. Second, the cost of growing cover crops may outweigh the benefits an individual farmer receives. Most covers give no income and this restricts their use on smallholdings where farmers do not have sufficient cash reserves to wait for the tree crop to mature. Research is required to find suitable cover crops, particularly varieties of beans and peas, which the smallholder can grow. Third, ground covers compete for the available moisture and, in dry areas, may adversely affect the growth of the main crop. Studies in rubber plantations in eastern Java show that cover crops may reduce the soil moisture by up to 50 per cent during the dry season compared with clean-weeding (Williams and Joseph, 1970). An alternative conservation measure is required in these circumstances.

6.1.3. Strip cropping

With strip cropping, row crops and protection-effective crops are grown in alternating strips aligned on the contour or perpendicular to the wind. Erosion is largely limited to the row-crop strips and soil removed from these is trapped in the next strip downslope or downwind which is generally planted with a leguminous or grass crop. Strip cropping is best suited to well-drained soils because the reduction in runoff velocity, if combined with a low rate of infiltration on a poorly drained soil, can result in waterlogging and standing water.

When used to protect land against water erosion, strip cropping is not normally required on slopes less than 3° and, on its own, is insufficient to conserve soil on slopes over 8.5°. Strip widths vary with the degree of erosion hazard but are generally between 15 and 45 m. Where possible, crop rotation is applied to the strips but, on steeper slopes or on very erodible soils, it may be necessary to retain some strips in permanent vegetation. These buffer strips are usually 2 to 4 m wide and are placed at 10 to 20 m intervals. The selection of strip widths and the design of strip layouts are based on practical experience. Guidelines for widths are given in F.A.O. (1965) and Cooke and Doornkamp (1974). Diagrams of typical layouts are found in Schwab, Frevert, Edminster and Barnes (1966).

The main disadvantage of strip cropping is the need to farm small areas. This limits the kind of machinery that can be operated and, therefore, the technique is not compatible with highly mechanized agricultural systems. Although this is a less relevant consideration on smallholdings, the difficulty here is that much land is taken up with less valuable protection-effective crops. Where grass is grown, it cannot be used for pasture until after harvest unless the row-crop areas are fenced to keep out the stock. Insect infestation and weed control are additional problems associated with strip cropping.

6.1.4. Mulching

Mulching is the covering of the soil with crop residues such as straw, maize stalks, palm fronds or standing stubble. The cover protects the soil from raindrop impact and reduces the velocity of runoff and wind. From the conservation viewpoint, a mulch simulates the effect of a plant cover. It is most useful as an alternative to cover crops in dry areas where insufficient rain prevents the establishment of a ground cover before the onset of heavy rain or strong winds, or where a cover crop competes for moisture with the main crop.

In semi-humid tropical areas, the side effects of a mulch in the forms of lower soil temperatures and increased soil moisture are beneficial and may increase the yields of coffee, banana and cocoa. Elsewhere the effects of mulching can be detrimental. In cool climates, the reduction in soil temperature shortens the growing season whilst in wet areas, higher soil moisture may induce gleying and anaerobic conditions. Other problems with mulching are that the mulch competes with the main crop for nitrogen as it decomposes, special equipment is required to plant crops beneath a mulch, and, with normal usage, weed growth is encouraged.

A mulch should cover 70 to 75 per cent of the soil surface. With straw, an application rate of 0.5 kg m^{-2} is sufficient to achieve this. A lesser covering does not adequately protect the soil whilst a greater covering suppresses plant growth. Denser mulches are sometimes used under tree crops, once these have reached maturity, and, in these cases, they may successfully control weeds. The effectiveness of mulching in reducing erosion is demonstrated by the field experiments of Borst and Woodburn (1942) who found that, on a silt–loam soil on a 7° slope, annual soil loss was 2.46 kg m^{-2} from uncultivated, bare land but only 0.11 kg m^{-2} on land covered with a straw mulch applied at 0.5 kg m^{-2}. Similar results have been obtained in laboratory studies for the same soil, slope and mulch conditions by Lattanzi, Meyer and Baumgardner (1974). These authors quote soil loss rates under simulated rainfall of 1.87 kg m^{-2} h^{-1} with no mulch and 0.31 kg m^{-2} h^{-1} with the mulch.

6.2. SOIL MANAGEMENT

The aims of sound soil management are to maintain the fertility and structure of the soil. Highly fertile soils result in high crop yields, good plant cover and, therefore, in conditions which minimize the erosive effects of raindrops, runoff and wind. These soils have a stable, usually granular, structure which does not break down under cultivation, and a high infiltration capacity. Soil fertility can thus be seen as the key to soil conservation.

The most practical way of achieving and maintaining a fertile soil is to apply organic matter. This improves the cohesiveness of the soil, increases its water retention capacity and promotes a stable aggregate structure. As seen in section 3.2, soils with less than 2 per cent organic content are generally erodible. To increase the resistance of an erodible soil by building up organic matter is a lengthy process, however, because organic content must be raised by 1 to 2 per cent before any effect on stability is observed.

Organic material may be added as green manures, straw or as a manure which has already undergone a high degree of fermentation. The effectiveness of these three kinds of material varies with the isohumic factor, which is the quantity of humus produced per unit of organic matter. Green manures, which are normally leguminous crops ploughed in, have a high rate of fermentation and yield a rapid increase in soil stability. The increase is short-lived, however, because of a low isohumic factor. Straw decomposes less rapidly and so takes longer to affect soil stability but has a higher isohumic factor. Previously fermented manures require still longer to influence soil stability but their effect is longer lasting because these have a still higher isohumic factor (Fournier, 1972).

The value of organic matter is enhanced by the presence of base minerals in the soil as these combine with the organic materials to form by chemical bonding the compounds of clay and humus which make up the soil aggregates. The base minerals are thus retained in the soil rather than removed by leaching or subsurface flow. Where these minerals, which provide the essential nutrients for plant growth, are absent, they should be added to the soil as fertilizer in the amounts normally recommended for the crop being grown. Mineral fertilizers on their own have no long term effect on the state of humus in the soil and they need organic support to bring about an improvement in aggregate structure. Moreover, the continual use of mineral fertilizers without organic manures may lead to structural deterioration of the soil and increased erodibility. Lime should also be applied to the soil as this reduces acidity and, by encouraging the growth of legumes, may ensure the success of their use in erosion control.

When managed so as to maintain their fertility, most soils retain their stability and are not adversely affected by standard tillage operations. Indeed, tillage is an essential management technique. It provides a suitable seed bed for plant growth and helps to control weeds. Problems arise on dusty, fine sandy soils, particularly when dry; on very heavy, sticky soils; and on structureless soils, especially those with a high sodium content.

In the first case, conventional tillage practice tends to pulverize the soil near the surface and create a compacted layer at plough depth which reduces infiltration and results in increased runoff. To overcome these effects tillage operations are restricted either by cutting down on their number by carrying out as many operations as possible in one pass, as with mulch tillage and minimum tillage, or by concentrating them only on the rows where the plant grows and leaving the inter-row areas untilled, as with strip-zone tillage (Schwab, Frevert, Edminster and Barnes, 1966; Table 6.1).

Reductions in runoff and therefore erosion on heavy soils are best achieved by increasing the rate of subsurface water movement by drainage. Erodible soils with more than 20 per cent clay content will benefit from the installation of mole drains and from the break up of compacted layers at depth by subsoiling.

Sodic soils are highly erodible because the excess of sodium results in the dispersal of the clay minerals on contact with water with consequent structural deterioration. Such soils appear to be particularly susceptible to tunnel erosion. The most effective treatment is to apply gypsum as this supplies a cation to replace the sodium. A good drainage system must also be provided to assist with washing out the sodium from the soil. This treatment is extremely expensive to employ over large areas, however, and, unless accompanied by ripping to break up the tunnels and the sowing of grass, gives only temporary relief.

Temporary stability, lasting from two weeks to six months, can be obtained on most soils by using soil stabilizers or soil conditioners. These oil or rubber based compounds, normally applied with water as a spray, are poly-functional polymers which develop chemical bonds with the minerals in the soil. They are too expensive for general agricultural use but, where the cost is warranted, are helpful on special sites like sand dunes, road cuttings or embankments and stream banks, to provide a short period of stability between the times of seeding and establishment of a plant cover. More permanent protection of banks and embankments can be achieved by facing them with a resistant material such as concrete. Better, however, is to construct retaining walls with gabions. These are rectangular steel-wire mesh baskets, packed tightly with stones, and they have the advantages of allowing seepage of water through the facing and of deforming by bending without loss of structural efficiency rather than by cracking.

6.3. MECHANICAL METHODS

Mechanical field practices are used to control the movement of water and wind over the soil surface. A range of techniques is available and the decision on which to adopt depends on whether the objective is to reduce the velocity of runoff and wind, increase surface water storage capacity or safely dispose of excess water. Mechanical methods are normally employed in conjunction with agronomic measures.

6.3.1. Contouring

Carrying out ploughing, planting and cultivation on the contour can reduce soil loss from sloping land by up to 50 per cent compared with cultivation up-and-down the slope. The effectiveness of contour farming varies with slope steepness and slope length, for it is inadequate as the sole conservation measure for lengths greater than 180 m at 1° steepness. The allowable length declines with increasing steepness to 30 m at 5.5° and 20 m at 8.5°. Moreover the technique is only effective during storms of low rainfall intensity. Protection against more extreme storms is improved by supplementing contour farming with strip cropping (section 6.1.3). The soil loss from contour strip-cropped fields is 25 to 45 per cent of that from fields managed by up-and-down tillage depending upon the slope steepness.

On silty and fine sandy soils, erosion may be further reduced by storing water on the surface rather than allowing it to run off. Limited increases in storage capacity can be

Table 6.1 Tillage practices used for soil conservation

PRACTICE	DESCRIPTION
Conventional	Standard practice of ploughing with disc or mouldboard plough, one or more disc harrowings, a spike-tooth harrowing, and surface planting.
Strip or zone tillage	Preparation of seedbed by conditioning the soil along narrow strips in and adjacent to the seed rows, leaving the intervening soil areas untilled: e.g. plough-plant; wheel-track planting; listing.
Mulch tillage	Practice that leaves a large percentage of residual material (leaves, stalks, crowns, roots) on or near the surface as a protective mulch.
Minimum tillage	Preparation of seedbed with minimal disturbance. Use of chemicals to kill existing vegetation, followed by tillage to open only a narrow seedband to receive the seed. Weed control by herbicides.

After Schwab, Frevert, Edminster and Barnes (1966).

obtained by forming ridges, usually at a slight gradient of about 1:400 to the contour, at regular intervals determined by the slope steepness. Greater storage is achieved by connecting the ridges with cross-ties over the intervening furrows, thereby forming a series of rectangular depressions which fill with water during rain. Because crop damage can occur if the water cannot soak into the soil within 48 h, this practice, known as tied-ridging, should only be used on well-drained soils. If it is applied to clay soils, waterlogging is likely to occur. A similar technique, called range-pitting, is sometimes used on grazing land whereby a series of pits is dug, the pits being about 50 cm by 50 cm, 7.5 cm deep and 40 cm apart.

6.3.2. Contour bunds

Contour bunds are earth banks, 1.5 to 2 m wide, thrown across the slope to act as a barrier to runoff, to form a water storage area on their upslope side and to break up a slope into segments shorter in length than is required to generate overland flow. They are suitable for slopes of 1° to 7° and are frequently used on smallholdings in the tropics where they form permanent buffers in a strip-cropping system, being planted with grasses or trees (Roose, 1966). The banks, spaced at 10 to 20 m intervals, are generally hand-constructed. There are no precise specifications for their design and deviations in their alignment of up to 10 per cent from the contour are permissible.

6.3.3. Terraces

Terraces are earth embankments constructed across the slope to intercept surface runoff and convey it to a stable outlet at a non-erosive velocity, and to shorten slope length. They thus perform similar functions to contour bunds. They differ from them by being larger and designed to more stringent specifications. Decisions are required on the spacing and length of the terraces, the location of terrace outlets, the gradient and dimensions of the terrace channel and the layout of the terrace system. The procedures for terrace design are described in Schwab, Frevert, Edminster and Barnes (1966) and Hudson (1971).

Terraces can be classified into three main types: diversion, retention and bench. The primary aim of diversion terraces is to intercept overland flow and channel it across the slope to a suitable outlet. They therefore run at a slight gradient, usually 1:250, to

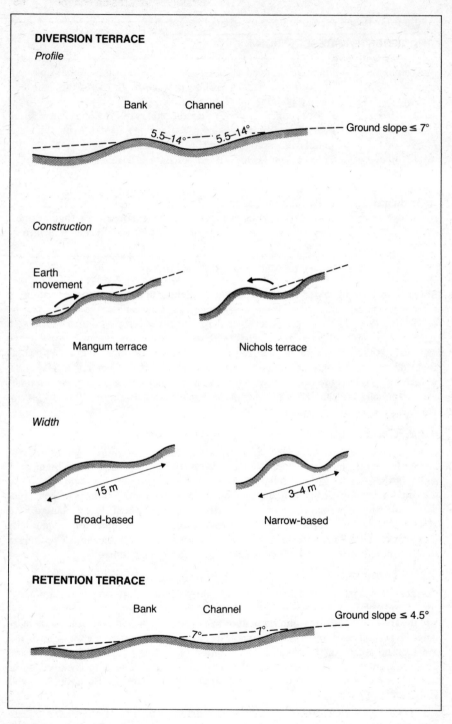

Fig. 6.1 Terraces.

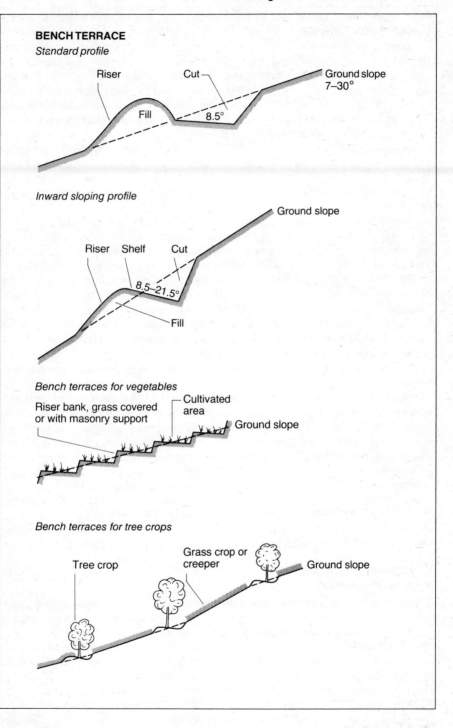

BENCH TERRACE
Standard profile

Riser   Cut   Ground slope 7–30°

Fill   8.5°

Inward sloping profile

Ground slope

Riser   Shelf   Cut

8.5–21.5°

Fill

Bench terraces for vegetables

Riser bank, grass covered or with masonry support   Cultivated area   Ground slope

Bench terraces for tree crops

Tree crop   Grass crop or creeper   Ground slope

Fig. 6.1 *(cont.)*

the contour. There are several varieties of diversion terrace. The Mangum terrace, formed by taking soil from both sides of the embankment, and the Nichols terrace, constructed by moving soil from the upslope side only, are broad-based with the embankment and channel occupying a width of about 15 m. Narrow-based terraces are only 3 to 4 m wide and consequently have steeper banks which cannot be cultivated. For cultivation to be possible, the banks should not exceed 14° slope if small machinery is used or 8.5° if large reaping machines are operated. Diversion terraces are not suitable for ground slopes greater than 7° because of the expense of construction and the close spacing which would be required.

Retention terraces are used where it is necessary to conserve water by storing it on the hillside. They are therefore ungraded or level and generally designed with the capacity to store runoff volume with a ten-year return period without overtopping. These terraces are normally recommended only for permeable soils on slopes of less than 4.5°.

Bench terraces consist of a series of alternating shelves and risers and are employed where steep slopes, up to 30°, need to be cultivated. The riser is vulnerable to erosion and is protected by a vegetation cover and sometimes faced with stones or concrete. There is no channel as such but a storage area is created by sloping the shelf into the hillside (Fig. 6.1). The basic bench terrace system can be modified according to the nature and value of the crops grown. Two kinds of system are used in Peninsular Malaysia (Williams and Joseph, 1970). Where tree crops are grown, the terraces are widely spaced, the shelves being wide enough for one row of plants, usually rubber or oil palm, and the long, relatively gentle riser banks being planted with grass or a ground creeper. With more valuable crops such as temperate vegetables grown in the highlands, the shelves are closely spaced and the steeply sloping risers frequently protected by masonry.

Although bench terraces appear to be reasonably satisfactory as a conservation measure in Malaysia, they are not necessarily suited to all steeply sloping land. In the Uluguru Mountains in Tanzania, they are unsuitable because, according to Temple (1972*b*), the topsoil is too thin so that their construction exposes the infertile subsoil, they require too high a labour input for construction and maintenance, and they hold back so much water on the hillsides that the soils become saturated and landsliding is induced. As an alternative conservation measure, the use of step or ladder terraces is recommended. These consist of narrow shelves, similar in size to terracettes, cut by a hoe. Little subsoil is dug out and during cultivation vegetation and crop residue are spread over the shelves and covered with soil cut from the face of the terrace upslope. Only rarely do ladder terraces break down under heavy rain (Temple and Murray-Rust, 1972).

6.3.4. Waterways

The purpose of waterways in a conservation system is to convey runoff at a non-erosive velocity to a suitable disposal point. A waterway must therefore be carefully designed. Normally its dimensions must provide sufficient capacity to confine the peak runoff from a storm with a ten-year return period. Three types of waterway can be incorporated in a complete surface water disposal system. These are diversion channels, terrace channels and grass waterways (Fig. 6.2). Diversions are placed upslope of areas of farmland to intercept water running off the slope above and divert it across the slope to a grass waterway. Terrace channels collect runoff from the inter-terraced areas and also convey it across the slope to a grass waterway. Grass waterways are therefore designed to transport downslope the runoff from these sources to empty into the natural river

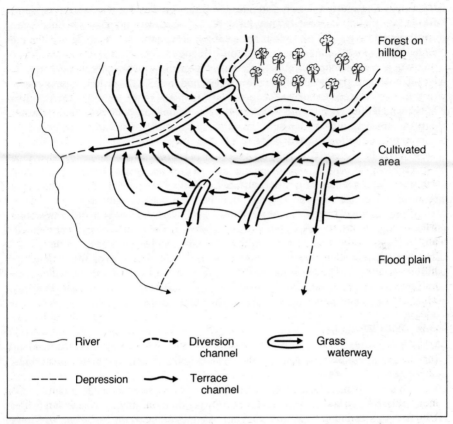

Fig. 6.2 Typical layout of waterways in a soil conservation scheme.

system. The grass waterways are located in natural depressions wherever possible. Occasionally, natural channels are reshaped to serve as grass waterways. Design procedures for waterways are found in Schwab, Frevert, Edminster and Barnes (1966) and Hudson (1971).

Grass waterways can be replaced in the water disposal system by tile drains. Diversion and terrace channels are graded to a soak-away, normally located in a natural depression, which provides the intake to the drain. The tile system is designed to remove surface water over a period not exceeding 48 h so that crop damage does not occur. Soil loss from tile-outlet terraces is much reduced because less than 5 per cent of the sediment delivered to the soak-aways passes into the drainage system (Laflen, Johnson and Reeve, 1972).

6.3.5. Stabilization structures

Stabilization structures play an important role in gully reclamation and gully erosion control. Small dams, usually 0.4 to 2.0 m in height, made from locally available materials such as earth, wooden planks, brushwood or loose rock, are built across gullies to trap sediment and thereby reduce channel depth and slope. These structures have a high risk of failure but provide temporary stability and are therefore used in association with agronomic treatment of the surrounding land where grasses, trees and shrubs are planted. If the agronomic measures successfully hold the soil and reduce runoff, the dams can be allowed to fall into disrepair. Even though they are temporary, the dams have to be carefully designed. They must be provided with a spillway to deal with overtopping during high flows and installed at a spacing appropriate to the slope of the channel. Design procedures are described in Heede (1976).

More permanent structures are sometimes required on large gullies to control the overfall of water on the headwall. These are designed to deal safely with the peak runoff with a ten-year return period. They must therefore dissipate the energy of the flow in a manner which protects both the structure and the channel downstream. The structures comprise three components: an inlet, a conduit and an outlet. The various types of each component are outlined in Schwab, Frevert, Edminster and Barnes (1966).

It should be stressed that gully erosion control is both difficult and expensive. One of the problems is correctly identifying the cause of erosion. This is extremely important because measures aimed at treating surface erosion will not always prevent and may aggravate subsurface erosion. Although the main approach of using agronomic measures to reduce runoff helps to control both kinds of erosion, it is in the supporting measures that differences exist. For surface erosion, the aim is to control water movement which, as seen above, is achieved by increasing depression storage behind small dams. For subsurface erosion, the aim is to obtain a uniform pattern of infiltration. The ponding of water works against this by providing concentrations of water which may be sufficient to open up a tunnel system. Where the tunnels have reached an advanced stage, they must be broken up by ripping prior to agronomic treatment.

6.3.6. Windbreaks

Shelterbelts are placed at right angles to erosive winds to reduce wind velocity and, by spacing them at regular intervals, break up the length of open wind blow. Shelterbelts are strictly living wind breaks. Inert structures such as stone walls, slat and brush fences and cloth screens can be used to perform similar functions on a smaller scale.

A shelterbelt is designed so that it rises sharply on the windward side and provides both a barrier and a filter to wind movement. A complete belt consists of two tree rows and up to three shrub rows, one of which is placed on the windward side. The density of the belt should not be so great as to form an impermeable barrier nor so sparse that the belt is transparent. A belt of the correct density is best described as translucent. Tree belts occupy a width of about 9 m and give effective protection over a distance equal to about twelve times the belt height. The length of the belt should be at least twenty-four times its height. Where hedges are used, the belt is about 3 m wide and the effective distance protected is about thirty times the belt height but, because of their lower height, the absolute distance protected by hedges is less than that for tree belts.

The plant species should be rapid growing, tolerant of wind and light, and frost resistant where necessary. Preference should be given to local rather than imported species. The side effects of shelterbelts are decreased evapotranspiration, higher soil temperatures in winter but lower in summer, and greater risk of weeds and pests. In the design of shelterbelts, decisions are required on belt height, spacing, width, length, shape and layout. Design procedures are found in Schwab, Frevert, Edminster and Barnes (1966).

6.4. CONSERVATION STRATEGY

Soil conservation strategies are aimed at reducing erosion to an acceptable level. Theoretically, this level is that at which the rates of soil loss and soil formation are balanced. However, in practice, it is difficult to recognize when this balanced state exists and, for this reason, an alternative definition of an acceptable level is adopted, namely that at which soil fertility can be maintained over 20 to 25 years. A mean annual soil loss of 1.3 kg m^{-2} is generally accepted as the maximum permissible but values as low as 0.2 to 0.5 kg m^{-2} are recommended for particularly sensitive areas where soils are thin or highly erodible (Hudson, 1971). Although these figures represent targets which should always be aimed at, they may be unrealistic for areas where erosion rates are naturally high, as in mountainous terrain with high rainfall. Under these conditions a target of 2.5 kg m^{-2} is more reasonable. All of these soil loss rates apply to field size units and they should be modified to take account of the greater likelihood of deposition of sediment within rather than loss from larger areas. Target values of 0.2 kg m^{-2} and 2.5 kg m^{-2} have been suggested for areas larger than 10 km^2 and smaller than 1 ha respectively (Morgan, in press).

When deciding what conservation measures to employ, preference is always given to agronomic treatments. These are less expensive and deal directly with reducing raindrop impact, increasing infiltration, reducing runoff volume and decreasing wind and water velocities. It is also easier to fit them into an existing farming system. Mechanical measures are largely ineffective on their own because they cannot prevent the detachment of soil particles. Their main role is in supplementing agronomic measures, being used to control the flow of any excess water and wind that arise. Many mechanical works are costly to instal and maintain. Some, such as terraces, create difficulties for farmers. Unless the soils are deep, terrace construction exposes the less fertile subsoils and may therefore result in lower crop yields. On irregular slopes, terraces are varied in width, making for inefficient use of farm machinery and only where slopes are straight in plan can this problem be overcome by parallel terrace layouts (Kohnke and Bertrand, 1959). Also, there is always a risk in severe storms with return periods in excess of twenty years of terrace failure. When this occurs, the sudden release of the water ponded up on the hillside can do more damage than if no terraces had been constructed. For all these reasons, terracing is often unpopular with farmers.

Although conservation schemes must be well designed if they are to effectively reduce erosion and not fail, their ultimate success depends on how well the measures are implemented. The willingness of farmers and others to adopt the techniques required by a particular strategy is thus as important as the ability of the engineer to design them. Equally important is that the strategy proposed clearly relates to the problems involved. Thus conservation design most logically follows a thorough assessment of erosion risk.

CHAPTER 7
RECONNAISSANCE
EROSION SURVEY IN
PENINSULAR MALAYSIA

The next three chapters are concerned with the practice of erosion survey for conservation design. They show how a picture of erosion risk and an understanding of the erosion
processes operating in an area can be built up using the procedures outlined in Chapter
4, some of the modelling techniques introduced in Chapter 5 and the basic coverage of
the mechanics of erosion in Chapters 2 and 3. Although Peninsular Malaysia is used as
the case study area, the approach adopted in these chapters can be easily applied
elsewhere. Among the advantages of choosing Peninsular Malaysia, however, are that it
has many of the erosion problems typical of tropical agriculture and that no previous
erosion surveys have been carried out there. Further, although information on climate,
soils and landforms is well documented, there are still some gaps in the data. These
provide the opportunity to illustrate possible approaches to erosion assessment when
data are scarce. There is also a general awareness of soil erosion as a problem as
witnessed by the widespread use of bench terracing and ground covers as conservation
measures, legislation such as the Silt Control Enactments of 1917 and 1922 and the
Land Conservation Act, 1970, and the report on soil and water conservation by Speer
(1963) commissioned by the Federal Government.

   This and the two following chapters (8 and 9) describe surveys carried out at
different scales for different purposes. This chapter covers a reconnaissance survey of
the whole country. Chapter 8 deals with a semi-detailed survey in central Pahang and
Chapter 9 describes a detailed survey of a small catchment and shows how it is used in
conservation design. These three scales of operation conform to the relationships
between erosion and scale identified in Chapters 1 and 4 (Table 1.1).

7.1. PENINSULAR MALAYSIA

Peninsular Malaysia is dominated by two landform associations: strongly dissected
highlands and low-lying coastal and riverine plains. The main mountain ranges, forming
a series of roughly parallel, north–south ridges, rising to over 2 000 m, have a landscape
of steep-sided, winding valleys, with boulder-strewn river beds, rapids and waterfalls.
Intrusive granitic rocks form the Banjaran Titiwangsa (Main Range) and Banjaran
Bintang, although remnants of a former sedimentary cover, remain in the former. The
Trengganu Plateau and Banjaran Timor, culminating in the peak of Gunung Tahan
(2 190 m), are composed of sedimentary strata. The mountains are flanked by low,
rounded hills, developed mainly on granites, or by areas of scarp-and-vale, etched out of
alternating sandstones and shales. The plains, occupying much of the west coast and the

lower parts of the major river valleys in the east, are composed of alluvium, most of it deposited following a post-Pleistocene rise in sea level. The plains extend far inland along the Pahang, Muar and Rompin valleys. Hillsides rise sharply from the flat land and the transition from low to high relief is very rapid. Deep weathering characterizes most of the country and regoliths up to 20 m or more in thickness are found on granitic rocks in the lowlands. Shallower mantles occur on sandstones and in most mountainous areas. Rock outcrops are rare, the most striking being in areas of tower karst and razor-back ridges of vein quartz.

The climate is equatorial monsoon. There are two main seasons: a southwest monsoon from April to September and a northeast monsoon from October to March. The mean daily maximum temperature in the lowlands is 32° C. Mean annual precipitation ranges from 1 750 to 2 510 mm. The average number of days with over 2 mm of rain is over 150 except for northern Perlis and along the Selangor coast where there is an average of at least one rain day in three. In spite of an apparent uniformity of climate over the country, there are important regional distinctions. The east coast has a strongly seasonal rainfall pattern with over 25 per cent of its annual rain falling in December and January. Because of this concentration of rain, intensities are higher and daily totals over 250 mm can be expected once in ten years. The rainfall over the rest of the country is more evenly distributed throughout the year but there are still significant variations in intensity. Daily totals with a ten-year return period range from 125 to 150 mm over most of the country but are less than 100 mm in the highlands (Morgan, 1971).

Numerous instances of soil erosion in Peninsular Malaysia have been documented, mainly in association with mining, agricultural and urban activities. References range from descriptions of high silt concentrations in rivers (Lake, 1894; Hartley, 1949) to attempts to quantify the amount of sediment being eroded (Fermor, 1939; Douglas, 1968b). Erosion has been observed as a result of gullying of logging tracks and roads in forested areas (Berry, 1956; Burgesss, 1971), gullying (Eyles, 1968a) and drain enlargement (Edgar, 1958) under rubber where inadequate ground cover is maintained, cultivation of land too steep for agriculture (Soper, 1938), and stream bank erosion in wet-rice growing areas (Maheswari, 1970). High rates of sediment loss occur in many mining and urban areas (Douglas, 1970).

Table 7.1 Measurements of soil loss in Peninsular Malaysia

LAND USE	RATE (kg m^{-2} y^{-1})	SOURCE
Rain forest	0.034	Shallow (1956) from R. Telom in Cameron Highlands.
Rain forest	0.004	Douglas (1972) from headwaters of R. Gombak
Tea plantation	0.673	Shallow (1956) from R. Bertam in Cameron Highlands.
Vegetables	1.009	Shallow (1956) from R. Kial in Cameron Highlands.
Mining	0.495	Douglas (1972) from R. Gombak, north of Kuala Lumpur.
Urban	0.800	Douglas (1972) from R. Anak Ayer Batu, near Kuala Lumpur.

Note: data from Douglas (1972) have been converted using a bulk density value of 1.0g cm^{-3}.

Few measurements of soil erosion have been made in the country and all the data available are derived from studies of sediment concentrations in rivers. Because much of the sediment removed from hillsides is deposited before it reaches the rivers, the data (Table 7.1) almost certainly underestimate the rates of soil loss. The values should be compared with a target of 0.2 kg m^{-2} as an acceptable level of soil loss (section 6.4). The data clearly show that erosion rates vary with the landuse. The highest rates are found in mining areas and in mountainous terrain used for agriculture. Extremely low rates characterize land under tropical rain forest.

These data do not permit more than the most generalized judgements on erosion risk in Peninsular Malaysia to be made. However, information is available on many of the factors which influence soil erosion and this can be used as a basis for assessing erosion risk. Meteorological data are obtainable from the Meteorological Service, the Drainage and Irrigation Department, schools, training colleges and universities, and rubber and oil palm estates. Details of the properties and distributions of soils are contained in the reports and maps of the Malaysian Soil Survey. Data on landforms, vegetation and landuse can be derived from topographical maps and aerial photographs, supplemented by the limited coverage of satellite imagery from LANDSAT. In addition, considerable use can be made of hydrological and geomorphological research. Studies of rainfall–runoff relationships under different landuses and of geomorphological processes operative in small drainage basins and on hillslopes are of particular value.

7.2. APPROACHES TO RECONNAISSANCE SURVEY

Two approaches are used for erosion assessment at the reconnaissance level. First, regional variations in various indices of erosion intensity are examined. Although, as indicated in Chapter 4, it is normally employed in semi-detailed surveys, precedents for this approach exist at this scale of survey. Stocking (1972) assesses erosion risk in Rhodesia on the basis of a stratified sample of gully density. Drainage density is used as an index of erosion severity in Bulgaria (Mikhailov, 1972) and Romania (Iana, 1972). The second approach employs rainfall data in an analysis of kinetic energy and rainfall aggressiveness along the lines described in sections 4.1.1 and 4.1.2.

7.2.1. Erosion intensity

Two measures of erosion intensity were tested: drainage density, defined as the length of streams per unit area, and drainage texture, defined as the number of first-order streams per unit area which, being equivalent to the density of source points, is analogous to gully density.

A map of drainage texture was derived as follows from a stratified random sample of third-order drainage basins. Within each 100×100 km grid square on the 1:63 360 topographical map sheets of the Malaysian Department of National Mapping, a single 1×1 km square was selected using random numbers as coordinates. If a third-order basin did not fall within or across the first square selected, additional squares up to twelve in number were examined in turn. If no basin was selected after twelve tries, that 100×100 km square remained unsampled. This was usually the case in areas of incoherent drainage patterns and swamp. It was assumed that the drainage pattern shown on the map sheets commences at the second order (Eyles, 1966) so that the number of first-order streams in each sampled basin can be estimated by multiplying the number of second-order steams by 3.5, which is the mean bifurcation ratio for Peninsular Malaysia (Eyles, 1968a). The average value of drainage texture was determined,

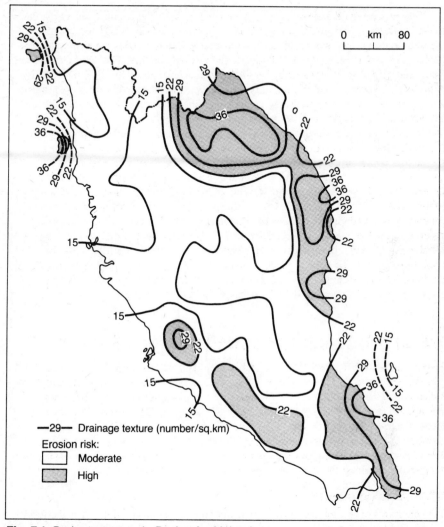

Fig. 7.1 Drainage texture in Peninsular Malaysia.

from a sample varying between 9 and 12 basins, for each of the 132 map sheets. These values were plotted on a map of Peninsular Malaysia and isolines drawn (Fig. 7.1).

A similar procedure was adopted for mapping drainage density (Fig. 7.2) but, in this case, use was made of the density values for 410 fourth-order basins obtained by Eyles (1968b). Average values, based on a sample of four or five basins, were obtained for each map sheet. For both texture and density, the number of drainage basins sampled represents about 3 per cent of their respective basin populations.

7.2.2. Rainfall analysis

The first method of analysis used the mean monthly rainfall totals for 680 recording stations compiled by the Drainage and Irrigation Department (1970) to produce a map of p^2/P (Fig. 7.3).

Next, an attempt was made to compile and map data on mean annual erosivity

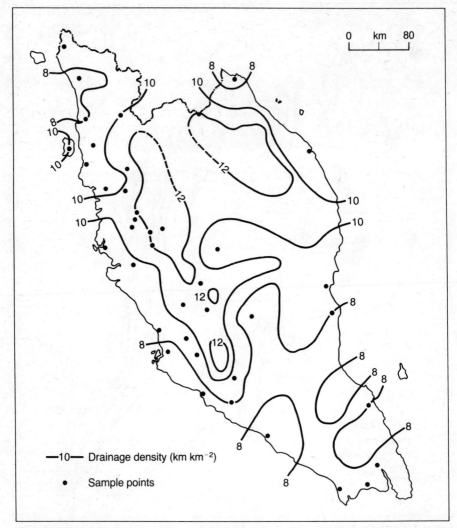

Fig. 7.2 Drainage density in Peninsular Malaysia.

using, for simplicity, the KE > 25 index. Insufficient information was available to obtain values for this index for more than a few stations. To help overcome this lack of data, the rainfall records for eleven stations of the Malaysian Meteorological Service with autographic rain gauges were examined for 1969 and a relationship sought between mean annual erosivity and a more widely available rainfall parameter. Taking only those days in that year which yielded over 50 mm of rain, the KE > 25 value was calculated for each day. Rainfall intensities were obtained directly from the autographic gauge charts and kinetic energy values were derived using equation (3.2) and the procedure set out in Table 3.4. Based on data for ninety-three days, the following regression equation was obtained:

$$EVd = 16.64Rd - 173.82 \qquad \text{with } r = 0.71, \tag{7.1}$$

where EVd is daily erosivity (J m^{-2}) and Rd is daily rainfall (mm).

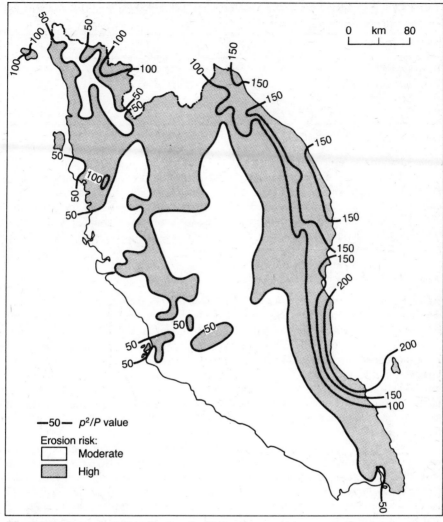

Fig. 7.3 Values of p^2/P in Peninsular Malaysia.

Separate equations were derived for the three rainfall regions of the lowlands (Morgan, 1971; Table 7.2). No equation could be obtained for the highlands because of insufficient data. Taking the values of the intercepts in the equations, it can be established that a daily rainfall total of 34 mm is required to register an erosivity value on the KE > 25 index in all regions. Using the equation appropriate to the location of each place, daily erosivity values were estimated for those days with 34 mm or more of rain, for ten recording stations for the years of 1965, 1966 and 1969. Daily rainfall totals were obtained from the monthly abstracts of the Malaysian Meteorological Service. Thirty values of annual erosivity were calculated by summing the daily values for each station for each year. It was found that these values (EVa) could be related to the mean annual rainfall (P; mm) by the following equation:

$$EVa = 9.28P - 8\ 838.15 \qquad r = 0.81. \tag{7.2}$$

Table 7.2 Regression equations for predicting daily erosivity from daily rainfall totals in Peninsular Malaysia

AREA	EQUATION		
Peninsular Malaysia	$EVd = 16.64Rd - 173.82$	$r = 0.71$	$n = 93$
West Coast climate	$EVd = 34.42Rd - 1\,121.97$	$r = 0.71$	$n = 23$
East Coast climate	$EVd = 16.16Rd - 357.17$	$r = 0.77$	$n = 45$
Port Dickson climate	$EVd = 26.06Rd - 553.85$	$r = 0.86$	$n = 16$

Equations from Morgan (1974); climatic types based on Morgan (1971).

This equation was used to estimate mean annual erosivity from mean annual rainfall totals, providing the basis for a map of erosivity (Fig. 7.4). It should be emphasized that although this approach to deriving erosivity values is universally valid, the resulting equations are not. Thus, whilst equation (7.2) appears to work well for Peninsular Malaysia, applying it to other countries is less satisfactory. With rainfall totals below 900 mm, the equation yields estimates of erosivity which are obviously nonsense.

7.3. IMPLICATIONS

Based on the work of Mikhailov (1972) and Iana (1972) arbitrary values of drainage density of 2 km km^{-2} and 15 km km^{-2} may be selected to represent moderate and severe erosion risks respectively. On this basis the whole of Peninsular Malaysia has a moderate risk of erosion and only in the highlands of the Trengganu Plateau and the Banjaran Titiwangsa does the risk become severe. The range of values for drainage density is low, however, compared to that for drainage texture which would seem to be the more sensitive parameter to regional variation. Taking an arbitrary value of 29 for drainage texture to separate areas of high and moderate erosion risk, high risk areas are found along the east coast, in western Johor, on Pulau Langkawi and Pulau Pinang and around Kuala Lumpur.

Values of p^2/P are highest in the eastern part of the country and in a belt on the west coast extending from Pulau Langkawi and Perlis in the north to Kuala Lumpur in the south. The belt extends eastwards into the mountain areas in the north but in the middle and south of the country lies to the west of them in the foothills. The highest erosivity values are found on the eastern side of the country and in a belt between Pulau Pinang and Kuala Lumpur in the west.

It has been shown elsewhere (Morgan, 1976) that, in Peninsular Malaysia, drainage density and drainage texture are virtually uncorrelated with each other and probably relate to different controlling mechanisms. High values of drainage density are associated with the transport of runoff from regular, moderate rainfalls whereas high values of texture are a response to a more seasonal rainfall regime with rains of greater intensity. Drainage density is better regarded as an index of runoff rather than erosion. In the same study, a significant correlation was observed between drainage texture and p^2/P ($r = 0.38$; $n = 39$). Both parameters are themselves significantly correlated with the daily rainfall total with a ten-year return period ($r = 0.57$; $n = 39$ and $r = 0.80$; $n = 21$ respectively). Because of the similarity between drainage texture and gully density and the association between high magnitude rainfall events and extension of the drainage system by the development of fresh first-order streams, as described by Nossin (1964) near Kuantan, both p^2/P and drainage texture may be regarded as indicators of the risk of gully erosion. In contrast, mean annual erosivity values reflect the risk of erosion by rainsplash, overland flow and rills (section 3.1.2).

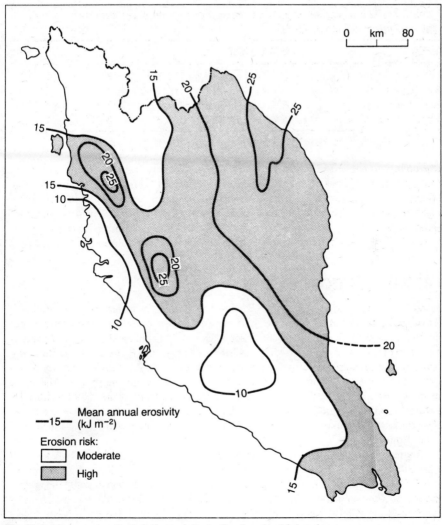

Fig. 7.4 Mean annual rainfall erosivity in Peninsular Malaysia.

It follows that by superimposing the maps of p^2/P and erosivity, a composite picture of erosion risk over the whole country is obtained (Fig. 7.5). Since there will always be at least a moderate risk of erosion in any tropical environment, the three categories of erosion risk recognized in this reconnaissance survey are designated severe, high and moderate, rather than introduce any category which is described as low.

The areas of severe erosion risk, defined as those where the risk of gully erosion and erosion by overland flow are both high, are found on the eastern side of the country, not only in the mountains of the Banjaran Timor and the Trengganu Plateau but also on the lower land of eastern Pahang and eastern Johor which include the areas occupied by the land settlement schemes of Pahang Tenggara and Johor Tenggara. The severe risk area extends westwards into the Banjaran Titiwangsa and then southwards, swinging to the west of the watershed into the foothills between Kampar and Kuala Lumpur. The area from Pulau Pinang eastwards to the southern part of the Banjaran Bintang around Taiping and Kuala Kangsar is also one of severe erosion risk.

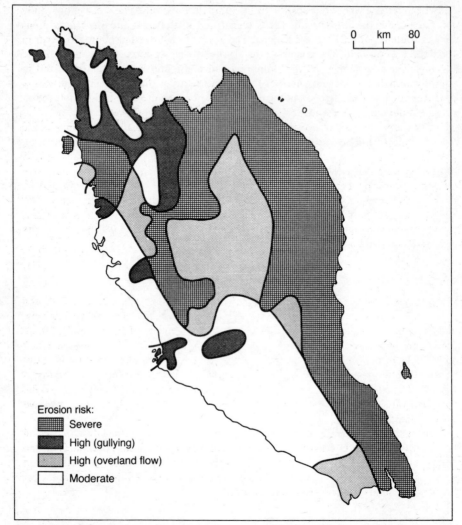

Fig. 7.5 Reconnaissance survey of soil erosion risk in Peninsular Malaysia.

Much of the land subject to severe risk is mountainous, steeply sloping and covered with rain forest. The role of the forest cover as a regulator of runoff and therefore a control over flooding has been stressed by Low and Goh (1972). It is therefore doubly important, for flood and erosion control, that this forest be preserved. Soil loss under rain forest is relatively low but increases rapidly when the land is cleared for agriculture (Table 7.1). This is true for the lower, hilly land falling into this category too. Great care is required in managing the development of all the areas with severe erosion risk. Large parts of the land are unsuitable for agriculture and should not be used in this way. It is doubtful whether they should even be exploited for timber. Wherever the land is found suitable for logging or for cultivation, the very best conservation practices must be adopted otherwise the dangers of flooding and erosion will be detrimental both locally and in the areas downstream.

The high risk areas are divisible into two groups: those where the main hazard is gullying and those where it is erosion by overland flow and rills. The areas of high gully risk are found in the northwest, in Perlis, Kedah and north Perak, and in scattered areas further south around Klang and Kajang. The areas of high overland flow risk occur in the middle Perak and Kinta valleys, the scarp-and-vale country to the north of the Pahang river and flanked by the Banjaran Titiwangsa and Banjaran Timor, and in southwestern Johor. Development of these areas requires that the best conservation practices are employed to combat the major erosion problem and that adequate measures are adopted to deal with the effects of the other erosive agents.

The moderate risk areas consist mainly of the coastal and riverine plains on the western side of the country with their associated hilly terrain. Sound conservation practices are necessary in the hilly areas but the risk on the plains comes more from outside than from any erosion problems generated within them. Where they lie downstream of areas of high and severe erosion risk, they are considerably affected by flooding and sedimentation resulting from poor erosion control in those areas.

An indication of actual rates of erosion can be obtained by using the equations relating mean annual sediment yield to p^2/P derived by Fournier (1960; Fig. 5.4). The following equations are appropriate to Peninsular Malaysia:

$$Qs = 27.12p^2/P - 475.40 \qquad \text{for lowlands} \qquad (7.3)$$
$$Qs = 52.49p^2/P - 513.20 \qquad \text{for highlands} \qquad (7.4)$$

where Qs is in t km^{-2} y^{-1}. These equations predict erosion rates between 0.88 and 2.23 kg m^{-2} y^{-1} for the lowland areas with p^2/P values between 50 and 100 respectively, and a rate of 2.11 kg m^{-2} y^{-1} in the highlands with a p^2/P value of 50. Comparing these predictions with measured rates of 0.495 and 0.80 kg m^{-2} y^{-1} in the lowlands and 0.67 and 1.09 kg m^{-2} y^{-1} in a part of the highlands where the p^2/P value is less than 50, it can be seen that they are of the right order of magnitude for areas affected by man-induced erosion. They are too high, however, for the rain-forested areas (Table 7.1). All the predicted values are in excess of 0.2 kg m^{-2}, the suggested target figure for acceptable soil loss (section 7.1).

Although reconnaissance surveys give a valuable general appreciation of erosion risk, they only indicate potential for erosion. The way in which this potential is turned into real erosion patterns depends upon the influence of factors other than climate on the erosion system. An analysis of soils, slopes and landuse is required and this means carrying out semi-detailed surveys of erosion.

CHAPTER 8
SEMI-DETAILED EROSION SURVEY IN CENTRAL PAHANG, PENINSULAR MALAYSIA

The objectives of semi-detailed erosion surveys are to identify the extent and type of soil erosion taking place in an area and to assess the relative importance of the various factors influencing soil loss. As indicated in section 4.2 it is possible to map the features of erosion such as rills and gullies but without being able to relate these to controlling variables, this approach is rather limiting. A survey is more meaningful if, in addition to showing the degree and type of erosion, information on relief, lithology, soils, slope and landuse is also portrayed. The survey then provides data for predicting soil loss and for formulating empirical models of the way in which the erosion system operates as well as serving as an inventory of information.

8.1. PROCEDURE

The mapping system of Williams and Morgan (1976), introduced in section 4.2.3, is adopted. In order to cover as large an area as rapidly as possible, most information is obtained by the interpretation of aerial photographs. Since most erosion features can be seen in direct stereoscopic image, the use of aerial photography can considerably reduce the cost per unit area of the survey compared with ground survey techniques. An indication of the extent of the reduction can be gauged from studies made of the costs of soil surveys. With only 15 per cent of the time spent on aerial photograph interpretation (Young, 1973), the saving on survey effort is over 50 per cent (Bie and Beckett, 1971).

Aerial photograph interpretation must be supported by fieldwork which allows the recognition of features on the photographs to be checked on the ground and the reproducibility of the interpretation to be assessed. A further and equally important aspect of fieldwork is to obtain data which cannot be derived from the photographs, particularly on soils and slopes.

The integration of aerial photograph interpretation and fieldwork is essential. For a semi-detailed survey the following procedure is recommended:

1 lay out the photographs or use satellite images, if available, as a photomosaic and draw boundaries between areas having distinctive photographic tones and patterns and which are therefore presumed to represent the boundaries between the major landform regions;
2 carry out stereoscopic examination of the photographs, plotting as much detail of the erosion features and related factors as possible;
3 carry out fieldwork to check this and obtain additional information;

4 carry out a second stereoscopic examination, finalizing the details to be mapped in the light of the fieldwork;
5 fair-draw the maps;
6 analyse and interpret the data;
7 assess or predict erosion risk; and
8 write report.

Although, as stated above, most erosion features are visible in direct stereoscopic image and are therefore identifiable by their form, shape and size, and their association with other features, the interpreter also makes use of tone, texture, shadow and pattern. The main erosion features mapped in this semi-detailed survey are: areas of rainsplash erosion and erosion by overland flow, rills, gullies and river bank erosion. Areas of soil wash by overland flow and splash erosion are recognized on black-and-white pan-chromatic photography as light grey tones, frequently in association with the top and middle slope segments. These light tones result from exposure of the subsoil, following removal of the top soil, and contrast with the darker tones of uneroded soils where the top soil with its organic content remains. Occasionally, where the parent rock and subsoil are dark in colour, this relationship between photographic tone and erosion is reversed (Bergsma, 1974).

The tonal patterns associated with splash and wash erosion are non-directional. Where the tones follow a linear pattern, especially with a downslope orientation, they are interpreted as rills. Under high magnification, a slight definition of depth to the rills may be apparent. Frequently rills show as an anastomosing pattern of light linearly arranged tones.

According to Brice (1966), a gully has steep sides, a steeply sloping or vertical headwall, a width greater than 0.3 m and a depth over 0.6 m. Gullies are thus distin-guishable on aerial photographs from rills by a clear definition of depth and from rivers by a channel with greater depth than width. Depending upon the quality and scale of the photography, and the degree of magnification used, width and depth measurements may be possible. Active gullies, usually carrying flow intermittently, have steep, unvege-tated banks, sometimes with evidence of slumping. Stable gullies have partially or even fully vegetated banks.

River bank erosion is characterized by the absence of vegetation, sometimes the presence of slumping, light tones and an association with the outside of bends. Addi-tional features which may be recognizable on aerial photographs are landslides, earth flows and areas of sedimentation.

8.2. STUDY AREA

The study area lies in the centre of Peninsular Malaysia, west of the River Pahang, between Jerantut in the north and Temerloh in the south. The rocks consist of alternat-ing shales and sandstones of the Kerdau Series, Triassic in age, and these have been moulded into smooth, steep-sided hills which rise abruptly from narrow valley floors. The local relief is low, only 100 to 120 m, but the dominantly convex slopes steepen rapidly from their crests to a maximum angle of 21°. At their lower end, a sharp break in slope angle occurs below which a piedmont slope grades gently into the valley floor (Fig. 8.1). There is a catenary sequence of soils with the Melaka Series occupying the hillslopes, the Durian Series on the lower slopes and the Batu Anam Series on the valley floors (Eyles, 1967). The mean annual rainfall is 2 052 mm and there is no dry season.

The return period of a daily rainfall of 50 mm is 118 days and the daily total with a ten-year return period is 186 mm. The hillslopes are covered in a mosaic of tropical rain forest, rubber plantations and secondary forest in various stages of growth. The valley floors are either used for wet rice cultivation or left as swamp.

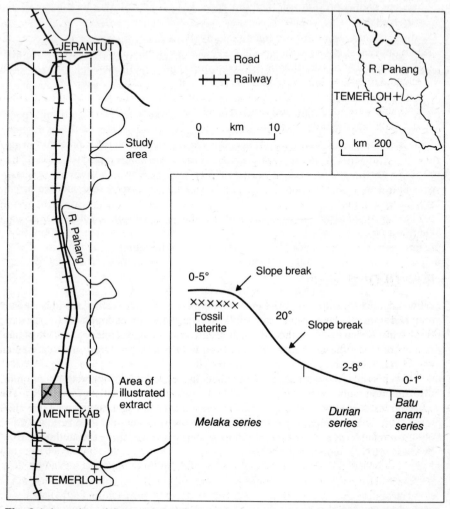

Fig. 8.1 Location of the semi-detailed erosion survey, central Pahang and detail of a typical slope profile in the area (after Eyles, 1967).

8.3. EROSION MAP

A soil erosion map of the area was produced from panchromatic aerial photography, flown in 1966 and with a nominal scale of 1:25 000, using a Wild ST4 Mirror Stereoscope with 3 × magnification and a Zeiss Interpretoscope with up to 15 × magnification. No satellite imagery was available so the photographs were used to form the photomosaic for the first stage of the work. The stereoscopic examination of the photographs proceeded as follows using the legend shown in Fig. 4.6 for plotting the information:

1 the study area was delimited on the photographs;
2 the stream pattern was plotted, separating permanent drainage lines from intermittent streams and dry valleys;
3 the crest pattern was plotted;
4 erosion features were identified and plotted, e.g. areas eroded by rainsplash and overland flow, rills, gullies, mass movements, river bank erosion;
5 areas of sedimentation were identified and plotted;
6 man-made features were shown, e.g. settlements and communications; areas affected by erosion due to runoff from badly located roads, poorly maintained terraces or inadequate waterways were marked;
7 areas where high runoff from adjacent uplands or convergence of surface flows is probable were identified and marked; and
8 the main landuse types were plotted.

The contour and drainage patterns shown on the 1:63 360 scale topographical map sheets covering the area were redrawn at 1:25 000 scale to provide a base on which to copy the information contained on the photographic plots. The whole was then fair-drawn at a scale of 1:10 000. An extract of the erosion map is shown in Fig. 8.2. It should be emphasized that the clarity of the erosion maps depends upon the use of colour and that reproduction in black and white, as here, can result in a somewhat cluttered effect.

8.4. SLOPE MAP

Although information on slope length and slope shape can be obtained from the erosion map and slope steepness can be calculated for specific localities from the contour data, there is insufficient room to show the spatial pattern of slope steepness on the same map. A map of slope classes, based on steepness, is therefore produced separately as an overlay (Fig. 8.3).

It is possible, though tedious, to produce the slope map from aerial photographs, using photogrammetric techniques, or to build up a map from field measurements. The latter are more appropriate for a detailed erosion survey over a small area (Ch. 9). A reasonably accurate map, sufficient for the purpose of a semi-detailed survey, can be obtained relatively easily and rapidly from topographic maps using the method devised by Olofin (1972). Taking the 1 × 1 km grid squares shown on the maps as a base, each square is divided into 0.14 × 0.14 km squares and, for these units, the maximum slope is estimated from the distance apart of the contours. The slope estimates are assigned to slope steepness classes according to the system developed by Olofin (1974) for use in Malaysia for planning purposes. In this system slope classes are based on morphological and theoretical ideas, relating slope angle and geomorphological processes, and on relationships between slope angle, limits to cultivation and the use of machinery. Detailed field checks by Olofin (1972) show that there is a very strong correlation between slope angles estimated by this method and those measured in the field with an Abney level.

8.5. FIELDWORK

The supporting field survey is designed to collect sufficient information to build up a realistic picture of the study area. Because of the impossibility of surveying the whole area in the field in a reasonable period of time, the survey is based on observations made

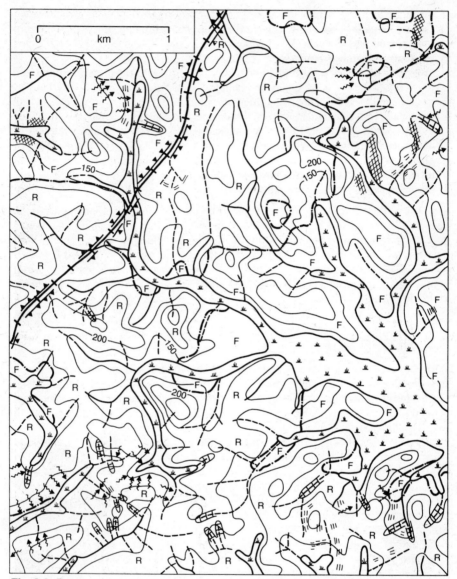

Fig. 8.2 Extract of the soil erosion survey map, central Pahang. Contour interval 50 ft.

at sample sites. The selection of these sites is critical. It must be made without bias, if their representativeness is to be assured, a condition which is essential if the results of the survey are to be extrapolated over the whole area.

8.5.1. Sampling design

The sampling procedure closely follows that recommended for selecting sites to monitor soil loss (section 5.1.1; Fig. 5.2). The first step is to delimit drainage basins of the third-order using the erosion map as a base. Third-order basins are chosen because they are of a size which confines them to similar lithological and relief conditions. The

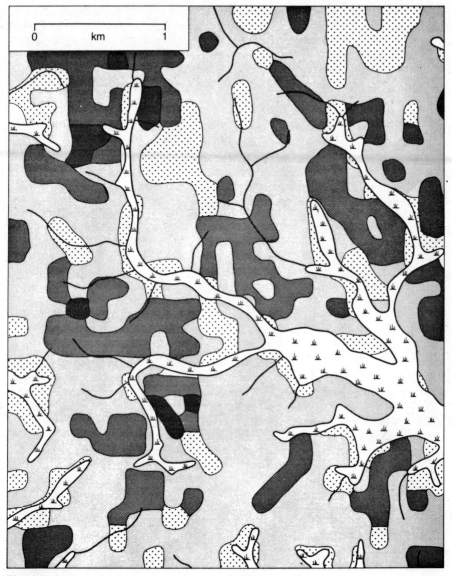

Fig. 8.3 Extract of the slope map, central Pahang.

drainage basins are given numbers and, using random number tables, a selection is made for field study.

The next stage is to select sample slope profiles within the drainage basins. A simple selection scheme is described by Young (1972). Each sample basin is treated separately and, using either a topographical map or aerial photographs, the divide and main drainage lines are plotted. The mid-slope line, which lies half-way between the divide and the drainage lines, is then drawn. Points along this line are selected either at random or at regular intervals and lines drawn from each point, along the steepest part of the slope, at right angles to the contours, to the divide and the drainage line. In practice, it is not always possible to survey all the transects selected because of inaccessibility. Where a transect cannot be surveyed, an alternative transect along a parallel line in close proximity may be substituted or it is omitted from the sampling scheme. In some cases, it is only possible to survey part of the transect.

Many observations on each transect are made at points and the selection of these is either at regular intervals or based on free survey, at the discretion of the surveyor. Normally, three points are used on each transect, located on the upper convexity, the mid-slope and the lower concavity.

The question of the representativeness of and the accuracy obtained from the sampling design is best approached at the point or site level. For each parameter being recorded, the required sample size can be determined which will result in a value for the sample mean which is within plus or minus a value (d) of the true mean, for a specified probability level (p). A preliminary sample of between ten and thirty observations is taken and the best estimate of the standard deviation ($\hat{s}$) from the mean is calculated from

$$\hat{s} = \sqrt{\frac{\Sigma (x - \bar{x})^2}{(n - 1)}} \tag{8.1}$$

where x is the value of any observation, $\bar{x}$ is the arithmetic mean of the series of observations and n is the number of observations in the preliminary sample. The sample size (N) for given values of d and p is obtained from

$$N = \frac{(\hat{s}t_{p,n-1})^2}{d} \tag{8.2}$$

where t is the value of Student's 't', obtained from tables, for probability level (p) and a preliminary sample size of n observations and $n - 1$ degrees of freedom. Reynolds (1975) has applied equation (8.2) to a study of soil properties and finds that to measure soil depth, pH, texture and organic content to an accuracy yielding a sample mean which differs from the true mean by the respective values of $\pm$ 5 cm, 0.1 units, 1 per cent and 5 per cent, at the 95 per cent probability level, requires a sampling density of 10 observations per 1 000 m². Clearly, standard soil survey procedures are less rigorous. Burrough, Beckett and Jarvis (1971) recommend a standard of five observations per cm² on the map. For a map at a scale of 1:20 000, 1 cm² = 40 000 m², which, according to the standard, requires 1 observation per 8 000 m². Generally, sampling density has to be adjusted to the time available for fieldwork.

8.5.2. Fieldwork procedure

Each slope transect is surveyed, beginning at either the divide or drainage line. Slope angles are measured to the nearest 1° with an Abney level, along successive sight lines, normally 10 m in length. Where the sight line is crossed by a marked break in slope, a shorter length is used, stopping the sight line at the slope break.

At the three selected sites on each transect, soil samples are taken with a 75-mm bore auger at 300 to 450 mm depth. In cultiviated areas this depth is usually sufficient to avoid disturbance of the top soil by terracing. The samples are removed to the laboratory for analysis of grain-size distribution, organic content and aggregate stability. Field tests may be made of infiltration capacity and shear strength using ¦ cylindrical infiltrometer and a portable shear vane respectively (Eyles, 1968c). Records of the plant cover, such as species, percentage ground cover, decaying vegetal matter and surface litter, are made within 1 × 1 m quadrats, centred on the sample points. Larger, 10 × 10 m quadrats are used to record bush and tree species and to analyse the

Recorder : Date : Altitude :		Area : Air photo no : Grid Reference :											FACET NO.
Present landuse													
Climate	Month	J	F	M	A	M	J	J	A	S	O	N	D	Erosivity
	Rainfall (mm)													
	Mean temp (°C)													
	Maximum intensity													
Vegetation	Type			% Ground cover						% Tree and shrub cover			
Slope	Position		Degree					Distance from crest		Shape			
Soil	Depth		Surface texture							Erodibility			
			Permeability			Clay fraction							
Erosion													
REMARKS													
EROSION CODE		0	½		1		2		3		4		5

Fig. 8.4 Pro-forma for recording soil erosion in the field (devised by Baker, personal communication).

structure of the plant association, particularly the number of layers involved and the extent of the canopy. Within the larger quadrat, an assessment is made of the state of erosion using a simple scoring system similar to that given in Table 4.4.

So that no items are forgotten, a standard booking procedure is adopted. A recording sheet or pro-forma (Fig. 8.4) is prepared, listing all the information required, and in the field an entry is made against each item even if only to denote its absence or that, for certain reasons, no observation was made.

Because detailed field studies of the type described above were carried out by Eyles

(1967; 1968b), no supporting field survey was conducted in this instance. Two recon-naissance visits were made to the area for checking purposes.

8.6. SOIL PROPERTIES

The soils have an inherently low erodibility. The topsoils of the Melaka, Durian and Batu Anam series have clay contents of 60 to 70 per cent (Ng, 1969), far higher than the upper limit associated with erodible soils (Evans, in press). The soils are therefore cohesive and resistant to detachment. Measurements of shear strength, at a depth of 0.3 m, and infiltration capacity made in the field (Eyles 1970; Table 8.1) show that soils of the Melaka Series are highly resistant to erosion. Their high shear strength means that they are not subject to slumping and the high infiltration capacity means that surface runoff is extremely rare. Soils of the Durian and Batu Anam Series are more susceptible to erosion. Rainfall intensities in excess of the infiltration capacity, and therefore surface runoff, are likely to occur as frequently as every 100 days on the Durian series and every month on the Batu Anam series.

8.7. RECONNAISSANCE ASSESSMENTS OF SEDIMENT YIELD

Taking the rainfall data for Temerloh as representative of the area, with a mean annual erosivity value of 10 204 J m^{-2} and a p^2/P value of 29, the risk of erosion can be classed as moderate (Figs. 7.3 and 7.4). Substituting the p^2/P value in equation (7.3) yields a mean annual erosion rate of 0.31 kg m^{-2} for small drainage basins. With the adoption of standard conservation practices based on agronomic measures supplemented by mechanical works on the steeper slopes it should be possible to reduce this to the acceptable level of 0.2 kg m^{-2}.

Table 8.1   Properties of soils near Kerdau, Pahang, Peninsular Malaysia

SOIL SERIES	SHEAR STRENGTH (kN m^{-2})	INFILTRATION CAPACITY (mm h^{-1})	NUMBER OF SAMPLES
Melaka	284.07	146.8	15
Durian	106.87	43.4	6
Batu Anam	143.42	12.2	5

After Eyles (1970).

8.8. ANALYSIS

A study of the erosion map (Fig. 8.2) reveals a close relationship between the severity of erosion and landuse. All the areas of accelerated erosion are confined to land cleared of the rain forest vegetation. Much eroded land has been temporarily abandoned and, at the time of the survey, was reverting to secondary forest.

Within these cleared areas, the location and type of erosion is determined by the interaction of slope form and soils. The summit areas are resistant to erosion, being covered by soils of the Melaka series. Erosion is here limited to rainsplash because overland flow and slumping rarely occur. The eroded areas are below the altitude of 45 m and are therefore concentrated almost entirely on the Durian series. The slope forms with their rapidly steepening convexity and lengths between 30 and 100 m are typical of erodible areas (Evans, in press). Where the slopes are straight in plan, erosion is by rainsplash, overland flow and rills. Where the slopes are concave in plan and there is moisture convergence towards the base of the slopes, gullies are found. Thus, gullying is restricted to the headwater areas and rill erosion to the valley sides. Occasionally, as in the area shown in the northeast of the extract, this erosion sequence in an individual drainage basin is completed downstream by deposition of sediment. Such a sequence emphasizes the importance of conservation in the headwaters for protection of the land downstream.

An index of soil erosion density was devised based on the product of the number and length of gullies per unit area. Data were collected from the erosion map for forty-two third-order drainage basins on: erosion density; drainage density, used as an index of runoff (section 7.3); relief; average maximum slope on the hillside above and contributing to the head of each first-order stream; average slope length between the head of each first-order stream and the divide; percentage area of the basin under tree crops, secondary forest and rain forest; relief ratio, defined as the difference in altitude between the highest and lowest points in the basin divided by the maximum basin length; and, as an index of basin shape, the lemniscate ratio, which relates maximum basin length (L) to basin area (A) in the form $L^2/4A$. Significant correlations exist between the soil erosion density index and drainage density ($r = 0.62$) and relief ($r = 0.33$). The best-fit regression equation, explaining 45 per cent of the variation in erosion density, is:

$$SED = 1.55Dd - 5.16K - 1.09, \tag{8.3}$$

where SED is soil erosion density (km km^{-2}), Dd is drainage density (km km^{-2}) and K is the lemniscate ratio (km km^{-2}). This equation and the above correlations imply that to reduce erosion requires a decrease in runoff and modifications to the form of the catchment to effectively lower the relief and make it more pear-shaped and less square or circular.

This empirical analysis and the predictions of sediment yield based on p^2/P (section 8.7) enable erosion risk to be assessed at a drainage basin scale. A semi-detailed erosion survey also permits the risk to be assessed at a hillslope scale. A simple way of carrying out this assessment is to predict soil loss using the following equation (Kirkby, (1976):

$$Qs = 170Qw^2 \tan \theta \tag{8.4}$$

where Qs is annual sediment yield per unit width from a 10 m slope length (cm^3 cm^{-1}), Qw is annual runoff (m^3 m^{-1}) and θ is slope angle. The annual runoff can be estimated from the mean annual rainfall using typical rainfall-runoff ratios of 0.25 for rain forest, 0.35 for rubber and 0.75 for urban areas (Low and Goh, 1972). For a rainfall total of 2 052 mm, these give respective runoff totals of 513, 718 and 1 539 mm. Assuming a unit area of 1 m^2 which, with a slope length of 10 m, discharges across a width of 0.1 m these values convert to 0.513, 0.718 and 1.539 m^3 m^{-1}. Predicted sediment yields for four slope angles are given in Table 8.2. If it is further assumed that the objective of soil conservation is to maintain soil loss at a rate equal to the rate of soil formation and that this rate is roughly equal to the erosion rate under natural vegetation, then, based on measurements by Douglas (1967a) in rain forest areas, a value of 0.2 kg m^{-2} y^{-1} represents an acceptable sediment yield. For a unit area, 10 m long and 0.1 m wide, and taking bulk density to be 1.0 g cm^{-3}, this value converts to 20 cm^3 cm^{-1} y^{-1}. Putting this value into equation (8.2) and predicting θ for the above runoff totals yields the maximum permissible slope angles at which acceptable sediment yields can be maintained. These are 3° for urban landuse, 13° for rubber and 24° for rain forest (Table 8.2). For rubber, conditions of mature trees and good ground cover are assumed.

Table 8.2 Predicted sediment yields near Kerdau, Pahang, Peninsular Malaysia

LANDUSE	Annual runoff ($m^3\ m^{-1}$)	**ANNUAL SEDIMENT YIELD** ($cm^3\ cm^{-1}$)				Maximum permissible slope*
		5°	10°	15°	20°	
Forest	0.513	3.91	7.88	11.99	16.28	24°
Rubber	0.718	7.66	15.45	23.48	31.89	13°
Urban	1.539	35.23	70.99	107.89	146.55	3°

Predictions based on slope length of 10m.
*Based on acceptable sediment yield of 20 cm^3 cm^{-1}.

8.9. IMPLICATIONS

In the light of the analysis of erosion risk, the following procedure for planning landuse and soil conservation strategies is recommended.

1 Using the slope map (Fig. 8.3) as a base, mark on an overlay the areas of land up to 14° and above 14° slope. This division represents a first shot at separating the land of lower slope, where the erosion risk from tree crop cultivation is acceptable, from that of steeper slope, where it is not.
2 Remove from the 0°–14° slope area all the valley floors, including the freshwater swamps, as these are too wet for tree crops.
3 Mark the drainage system on the overlay and, using the Strahler system of stream ordering (Gregory and Walling, 1973), delimit the third-order drainage basins.
4 Remove from the 0°–14° slope area all the headwater areas in square or circular third-order basins as having too great a risk of gully erosion to permit cultivation.
5 Add to the 0°–14° slope area, all straight valley side slopes between 15° and 19°, where tree crop cultivation is permissible if the relief of the land can be effectively reduced. It should be noted here that the legal upper slope limit for cultivation in Malaysia is 18.5°.
6 Three main landuse regions are now shown on the overlay, namely swampy valley floors, areas suitable for tree crops, and areas to be maintained under rain forest. Conservation measures can be designed for each region (Fig. 8.5).

The areas assigned to rain forest must be kept under dense vegetation to minimize erosion. They comprise steeply sloping land and areas of gentler slope with flow convergence on the more erodible soils of the Durian series. It may be possible to carry out well-managed logging operations using selective felling on land up to 24° slope but these should be undertaken only with the best erosion control. The natural erosion rate on land steeper than 24° is higher than the acceptable target value and this land should be protected as a watershed conservation measure. Attempts should be made to control erosion in the gullies shown on the erosion map by planting ground covers and shrubs and placing temporary dams across the channels.

In the areas devoted to tree crops it is essential that a dense ground cover is maintained. Any of the ground creepers mentioned in section 6.1.2 is suitable. In

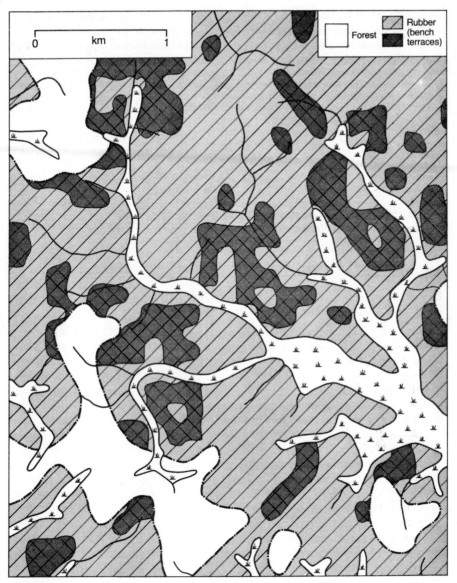

Fig. 8.5 Proposed landuse, central Pahang.

addition, bench terracing should be installed on slopes between 13° and 19° to reduce the relief available for surface runoff development by shortening slope length. Normally, bench terracing is recommended in tropical areas on slopes above 7° (Sheng, 1972a). In order to give maximum protection against erosion and enable a reasonably continuous network of terraces to be laid out, it is advisable to bench terrace all land between 9° and 19° slope. The terraces should be graded to natural depressions on the hillsides which drain downslope to the rivers and freshwater swamps.

Few conservation measures are necessary on the valley floors although stream banks liable to erosion should be stabilized. The adoption of the measures recommended in the other areas will prevent the valley floors from being subjected to excessive sedimentation.

The recommendations for soil conservation accompanying a semi-detailed erosion survey, though specific, require detailed design work before they can be implemented. The information needed for this is obtained from detailed surveys.

CHAPTER 9
DETAILED EROSION SURVEY
AND CONSERVATION STRATEGY

The purpose of a detailed erosion survey is to examine the feasibility of implementing the conservation strategy defined by a semi-detailed erosion survey and to provide a data base for the design of mechanical protection works. The survey described in this chapter is for a small catchment on the University of Malaya campus near Kuala Lumpur, Peninsular Malaysia. It was carried out in order to test detailed survey techniques.

9.1. STUDY AREA

The catchment occupies an area of 0.09 km² in the north of the campus. The bedrock consists of alternating sandstones and shales of the Kenny Hill Formation, interlensing in irregular sequence. The soil pattern is complex, reflecting the geology. Soils derived

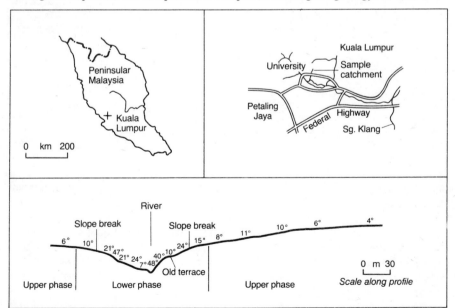

Fig. 9.1 Location of the detailed erosion survey, University of Malaya, Kuala Lumpur and detail of a typical slope profile (after Morgan, 1972).

from the sandstones have about 35 per cent clay and those on the shales over 40 per cent clay. The catchment has a relative relief of 61 m and slopes of a two-phase form (Fig. 9.1). The upper convexo-concave phase has maximum slopes of 19° to 25° and the lower phase comprises rectilinear segments of 30° to 48°. Sharp breaks, across which the slope angles differ by 10° or more, mark the boundary between the two phases and also the junction between the lower phase and a narrow flood plain. The catchment is drained by a second-order stream fed by three first-order gullies (Fig. 9.2).

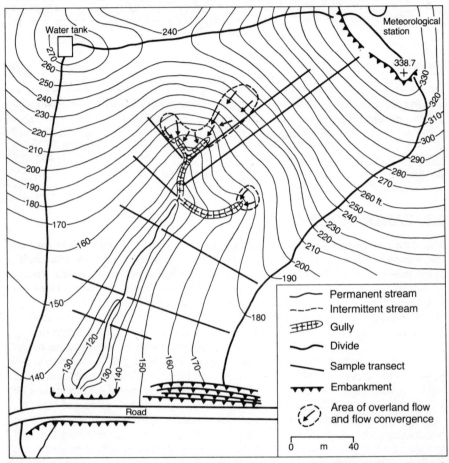

Fig. 9.2 Soil erosion map, study catchment, University of Malaya. Contour interval 10 ft.

The mean annual rainfall is 2 695 mm. The rainfall distribution is bimodal with highest monthly totals in April and October. The average number of storm days, defined as daily rainfall totals over 50 mm, is highest in April and December (Table 9.1). The return period of 50 mm in a day is 31 days and a daily rainfall total of 145 mm has a ten-year return period.

At the time of survey in 1969 and 1970, the eastern part of the catchment formed part of the field area of the Faculty of Agriculture and was planted with oil palm up to a height of 60 m and with rubber above that altitude. As a conservation measure the

Table 9.1   Rainfall data for the University of Malaya

	J	F	M	A	M	J	J
Mean monthly total (mm)	185	119	233	328	272	152	185
Mean days/month with over 50mm	0.8	0.3	1.0	1.3	1.0	0.7	1.0
	A	S	O	N	D		
Mean monthly total (mm)	198	178	307	277	262		
Mean days/month with over 50mm	1.0	0.6	0.9	0.7	1.3		

Data from the Department of Geography, University of Malaya, Meteorological Station, 1963–69.

slopes had been cut with 1 m wide bench terraces and a leguminous cover crop grown on the inter-terrace areas. The western part was occupied by the Science Faculty and was covered with reverted rubber and *belukar* or secondary growth.

9.2. LANDUSE PLAN

A landuse plan was simulated by applying the following procedure based on that adopted in semi-detailed erosion surveys. First, the objective was set to devote as much land as possible to tree crops without causing unacceptable erosion risk. Second, a mean annual sediment yield of 0.2 kg m^{-2} was selected as the maximum acceptable. Third, the maximum permissible slope angles at which this level of sediment yield can be maintained were determined using equation (8.4). Annual runoff values were estimated for the annual rainfall of 2 695 mm, using the rainfall-runoff ratios quoted in section 8.8. The results of this analysis show that rubber can be grown with the use of ground covers on slopes up to 7° and that the maximum acceptable sediment yield is exceeded under rain forest on slopes over 14°. It is recommended therefore that any extension of the rubber growing area on to steeper land using bench terracing should not incorporate all land up to the legal cultivable limit of 18.5° but should be restricted to slopes of 7° to 14°.

Clearly, this simulated plan envisages a considerably different landuse pattern from that existing at the time of survey and implies that some of the land in the catchment is being misused. It should also be noted that the maximum permissible slope angles are much lower than those prescribed for central Pahang because of the higher rainfall.

9.3. FACTORS AFFECTING FEASIBILITY

Our concern here is solely with those factors which affect the implementation and likely success of the conservation strategy. The wider issues of feasibility which affect agricultural plans generally are beyond the scope of this book. One approach to assessing the feasibility of a particular scheme is to pose a series of questions of which the following may serve as examples.

1 Does the size of the net farm area constitute a viable economic holding? The net farm

area is that part of the holding devoted to activities yielding an income either as revenue or in kind. Thus, on a rubber estate it would include the land planted with rubber trees but exclude land kept under forest and that taken up with access roads, waterways and buildings.

2 Is the net farm area sufficiently continuous that it provides a logical working unit? Where the area comprises several parts, is each part large enough to form a suitable working unit? Where a landuse proposal results in small disconnected units these must either be combined by farming the intervening areas which would not otherwise be cultivated or, where the erosion risk is too great to permit this, be taken out of cultivation.

3 Is there easy access to each part of the net farm area so that any necessary machinery can be brought in and produce removed?

4 Where terraces and waterways are planned, is the terraced area large enough for an efficient terrace and water-disposal system to be installed?

5 How well do the proposed conservation measures fit in with existing agricultural practices? If new measures are recommended, what evidence is there that these are economically and socially acceptable? What is the effect of the measures on crop yield?

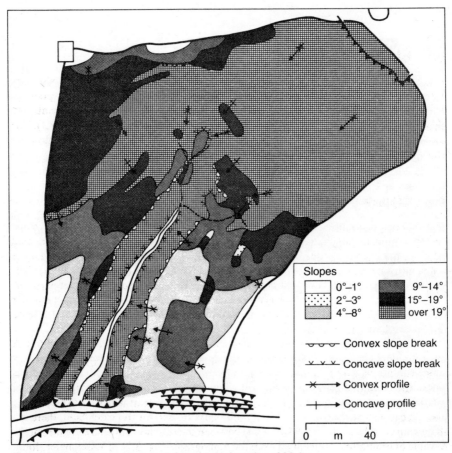

Fig. 9.3 Slope map, study catchment, University of Malaya.

6 Is the system legal? The main constraints here are to ensure that laws specifying upper slope limits to cultivation are adhered to and that no runoff is channelled on to another person's land which would not flow there naturally.

In the light of factors such as these, modifications are frequently necessary to the original proposals in order to design a strategy which stands a good chance of being successfully adopted.

9.4. EROSION SURVEY

The detailed erosion survey is intended to supplement the information obtained from a semi-detailed survey. Since most of the information for the latter derives from aerial photograph interpretation, additional data must be obtained largely by fieldwork. Fieldwork procedures closely follow those described in Section 8.5.

Slopes were measured along ten sample transects (Fig. 9.2) and soil samples were taken at three points on each transect corresponding to the convex segment on the upper slope phase, the junction between the upper and lower slope phases and the mid-slope segment of the lower slope phase (Fig. 9.1). Because some sites had to be abandoned through difficulty of access, soils were sampled at 24 sites only, giving a density of observations of 1 per 3 750 m^2 (section 8.5.1).

9.5. SLOPE MAP

A slope map was compiled from the field measurements and a 1:1 584 scale plan, contoured at 10-ft (3.048-m) intervals, prepared by the Estates Office, University of Malaya. On this map are shown the slope angles of the individual slope segments, the slope shape in profile and the location of slope breaks greater than 10° (Fig. 9.3). The main gullies, the areas of flow convergence and the land subject to overland flow are shown in Fig. 9.2.

9.6. SOILS

The soils belong to the Serdang Series. Soil properties vary locally with lithology, slope steepness and distance from the crest. The relationship between clay content and lithology has been outlined above. Silt content is higher on the steeper slopes and there is a significant correlation between silt content and slope angle ($r = 0.71; n = 24$). This implies that an increase in steepness promotes the selective erosion of the finer particles. Silt/clay ratios increase with distance from the crest, ranging from 0.64 on the upper phase, to 0.80 at the boundary between the upper and lower slope phases, and 1.33 on the lower phase (Table 9.2).

The silt/clay ratio may be used as an index of weathering (van Wambeke, 1962). A low ratio denotes a well-weathered soil, the development of which requires relatively stable conditions with minimal erosion. High ratios are associated with erodible areas where the continual removal of the soil allows insufficient time for a high degree of weathering to occur. Since the silt/clay ratio reflects the rate at which erosion is taking place, it can also be used as an index of erodibility. In fact, for this catchment, there is a significant correlation ($r = 0.95; n = 24$) between the silt/clay ratio and estimates of the K-value of erodibility determined from the nomograph (Fig. 3.2) of Wischmeier, Johnson and Cross (1971).

Table 9.2 Soil properties in a small catchment, University of Malaya

SITE	CLAY %	SILT %	SILT/CLAY RATIO	K VALUE (*)
Lower slope phase (topsoil)	35	47	1.33	0.17
Phase boundary (topsoil)	33	26	0.80	0.14
Upper slope phase (topsoil)	36	23	0.64	0.13
Upper slope phase (subsoil)	25	25	1.00	0.25
Upper slope phase (weathered bedrock)	20	42	2.10	0.31

(*) based on Figure 3.2.

Soil properties also change with depth, the clay content decreasing to 25 per cent in the subsoil and 20 per cent in the weathered bedrock. The silt/clay ratio increases with depth at first, reaching a miximum value at about 45 cm, before decreasing with greater depth (Fig. 9.4).

9.7. EROSION PROCESSES

The catchment was observed regularly in the field between September 1969 and April 1970 to monitor the frequency of overland flow and gully erosion. Observations were made at 09.00 h local time every Monday, Wednesday and Friday, and at other times during or immediately following heavy rain. Precipitation recordings were made at the Meteorological Station of the University of Malaya, sited on the divide in the northeast of the catchment. Although many occurrences of flow over the hillsides and in gullies were missed by not installing automatic monitoring equipment, sufficient data were obtained to provide an indication of the frequency and magnitude of rainfall events which cause erosion.

Overland flow was observed on three occasions, 16 October 1969, 30 and 31 March 1970, following the receipt of 16 mm or more of rain within an hour. It was never observed over the whole catchment but was restricted to those slopes adjacent to and above the gully heads. The flow was seen to converge downslope towards the gullies, pouring over their lips and down their backwalls before continuing downstream within the confines of the gully channels. On 16 October 1969, overland flow occurred for a distance of 60 m upslope of the gullies. Because of the dense vegetation cover, the overland flow did not cause erosion.

Channel flow in the gullies was observed on 18 occasions but only one event, that of 16 October 1969, when 73 mm of rain fell in an hour, resulted in erosion. Following this storm, it was found that scouring had locally increased the depths of the gullies by 30 to 45 cm and that, in one place, bank collapse had reduced the width of the gully by 2 m. Although nine occurrences of over 25 mm of rain in an hour were recorded in the study period, including one with an hourly total of 54 mm, no further erosion took place. It is concluded elsewhere (Morgan, 1972) that a rainfall total in excess of 60 m h^{-1}, or possibly even 75 mm h^{-1}, is required to contribute sufficient overland flow to the gullies to cause erosion. These events have a return period of about 60 days in this catchment.

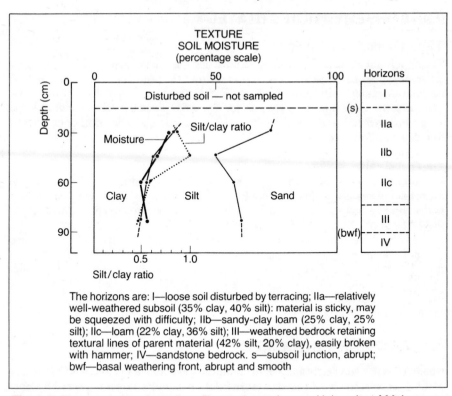

The horizons are: I—loose soil disturbed by terracing; IIa—relatively well-weathered subsoil (35% clay, 40% silt): material is sticky, may be squeezed with difficulty; IIb—sandy-clay loam (25% clay, 25% silt); IIc—loam (22% clay, 36% silt); III—weathered bedrock retaining textural lines of parent material (42% silt, 20% clay), easily broken with hammer; IV—sandstone bedrock. s—subsoil junction, abrupt; bwf—basal weathering front, abrupt and smooth

Fig. 9.4 Characteristics of a soil profile, study catchment, University of Malaya.

9.8. EROSION RISK

The mean annual erosivity, estimated from the mean annual rainfall, using equation 7.2, is 16 171 J m^{-2}. The p^2/P value is 39. These figures are indicative of a high erosion risk by overland flow and a moderate risk of gully erosion (section 7.3). This is in line with the process studies which reveal that most gully erosion results from the concentration of overland flow. The clay contents and K-values of the topsoil imply that erodibility is relatively low. Once the topsoil is broken through, as in the gully channels, the subsoil is highly erodible. The least erodible soils are found on the upper slope phase and erodibility increases rapidly downslope of the phase boundary. The most erodible area is clearly the valley head where slopes of comparable steepness to those of the lower phase occur in an area of flow convergence.

The catchment can be divided into three areas according to the degree of erosion risk, namely the valley head, the lower valley side slopes and the upper valley side slopes. Greatest risk exists in the valley head where overland flow and gully erosion are already taking place. Any decrease in plant cover will aggravate the erosive condition and the area should be classed as a conservation zone and maintained with a dense vegetation cover. The valley side slopes of the lower phase have a potential erosion risk and should not therefore be cultivated. The valley side slopes of the upper phase possess stable soils with relatively low erodibility and gentle slope angles. The main risk is from erosion by rainsplash and this can be counteracted by maintaining a dense ground cover if the area is to be cultivated.

9.9. CONSERVATION STRATEGY

9.9.1. Landuse allocation

Taking account of the erosion risk assessment and the maximum permissible slopes for different types of landuse, as recommended from the semi-detailed erosion survey, landuse allocation is straightforward. All land on the upper slope phase of the valley sides up to 14° slope may be safely used for tree crops. Land between 7° and 14° slope should be bench terraced. The rest of the catchment should remain under or be allowed to revert to a forest cover.

9.9.2. Forested areas

No special treatment of the hillslopes is required but a dense forest cover must be maintained. Because erosion is naturally high on slopes as low as 15° the forest should not be worked commercially for timber. This recommendation applies particularly to the valley head which is the most erodible area.

Erosion control measures are required in the gullies but because the area has no commercial value, high-cost permanent structures are not justified. Reliance must be placed on the protection afforded by the forest cover on the adjacent hillslopes supplemented by temporary dams. Given a gully depth of 0.5 m, wire-bound, loose-rock dams, 0.3 m high, are recommended. The spacing of the dams can be determined from the formula of Heede (1976):

$$\text{SPACING} = \frac{HE}{K \tan \theta \cos \theta} \tag{9.3}$$

where HE is the dam height and K is a constant. $K = 0.3$ for $\tan \theta \leq 0.2$; $K = 0.5$ for $\tan \theta > 0.2$. Dam spacings would need to range from 1.8 m on 19° slopes to 2.9 m 12° slopes and 10.4 m on slopes of 5° 30'. Dam construction should follow the guidelines laid down in Heede (1976).

9.9.3. Tree crop areas

The areas where tree crops may be safely grown occur on the upper valley sides in the east and west of the catchment and on the divide in the north (Fig. 9.5). The latter area suffers from poor access and will not be considered further. It is important that in the tree-crop growing areas a dense ground cover is provided to minimize the impact of raindrops and water drops falling from the canopy on the soil surface. Land clearance should be undertaken in June and July when erosivity is relatively low and ground creepers planted immediately so that they become established during August and September before the erosive rains in October. The most suitable creepers are *Pueraria phaseoloides, Calopogonium mucunoides* and *Centrosema pubescens* (Williams and Joseph, 1970). The ground cover must be maintained throughout the life of the trees.

Bench terracing must be constructed on land over 7° and on land of gentler slope where this is necessary to obtain a continuous terrace system. Construction should be carried out in late June or early July following land clearance. The modified bench-terrace system for tree crops (Fig. 6.1) should be adopted and the land between the terraces planted immediately with a ground creeper. Benches about 2 m wide are sufficient for tree crops. Using the spacing formula for bench terraces in Sheng (1972b) gives vertical interval spacings of 0.405 m on a 10° slope and 0.228 m on a 6° slope, corresponding to distances along the slope of 2.3 m and 2.18 m respectively. The wider spacing required on the steeper slopes reflects that fewer terraces are permissible in areas of greater erosion risk. These spacings are unsuitable for rubber cultivation,

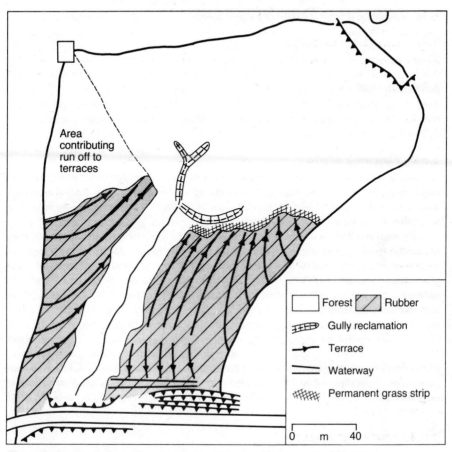

Fig. 9.5 Conservation scheme, study catchment, University of Malaya.

however, as the trees need to be placed about 9 m apart in one direction and 2.5 m in the other. The cheapest way of arriving at this spacing is to build the terraces 9 m apart and plant the trees at intervals of 2.5 m on each terrace.

The terraces should be located so that marked convex breaks of slope can be incorporated in the riser and concave breaks in the shelf. This avoids the problem of having sharp changes in slope, which are localities of higher erosion risk, often requiring extra protection, on land between the terraces. The terraces should be graded to a slope of 1:250 (Hudson, 1971) and their length kept below 100 m (Sheng, 1972b). On the western side of the catchment, the terraces can begin at the divide and be graded northeastwards. On the eastern side it is convenient to use a slight spur in the middle of the terraced area and grade the terraces both north and south of it (Fig. 9.5).

The accompanying waterway network is relatively simple. Only one area, in the west, could contribute runoff to the cultivated area and, because of its small size and the apparent lack of overland flow in that part of the catchment, it can be ignored. No diversion channels are therefore required. The terraces draining northwards can be allowed to empty into the forested area without the need for a waterway but a grass strip should be maintained between the end of the terraces and the forest to reduce the velocity of any runoff and filter out the sediment. A grass waterway is required for the terraces draining southwards on the eastern side of the catchment to intercept the runoff and prevent it cascading down the embankment above the road. This waterway should be designed according to procedures described in Schwab, Frevert, Edminster and Barnes (1966).

9.9.4. Assessment

Two changes are proposed to the existing landuse. The valley head, at present given over to rubber as part of the Faculty of Agriculture field area, should be taken out of cultivation and allowed to revert to secondary forest. The land in the west of the catchment which is currently under secondary growth is suitable for tree crops and could be cleared and planted with rubber.

The conservation scheme is feasible only if the catchment is considered as part of a larger area. The hectarage of the two areas where rubber is the proposed landuse is too small to comprise a viable economic unit even for a smallholder. Individually, however, the areas are large enough for working units within a larger holding. For the proposals to be realistic, the areas assigned to rubber would need to be extended, along with their associated terrace systems, across the divides on to similar land in the neighbouring catchments. Such extension is possible and, at the time of survey, had already taken place across the eastern divide within the Faculty of Agriculture field area. Access by tappers to the rubber areas is possible from the main road by climbing the divide and walking from there along the terraces. There is no reason to suppose that the recommended conservation practices are unacceptable to farmers because they are the standard measures employed in rubber-growing areas throughout the country. The proposals comply with legal requirements. Thus, even though the survey is unrealistic because of the small size of the area involved, it has enabled the techniques of detailed erosion surveys to be established and illustrated the potential application of such surveys over larger areas.

CHAPTER 10
CONCLUSION

In the last three chapters a hierarchical system of erosion survey at different scales has been presented. The objectives of the system are to further the understanding of the processes of erosion and how they operate in time and space and to furnish data which will lead to better conservation design. Although the general principles and working methods of the system have been established, the details are still being perfected. The case study of Peninsular Malaysia is one example of the application of the system and illustrates the current state of the art. It is intended that the system be usable in other environments even though modifications may be necessary to fit local conditions.

The justification of carrying out erosion surveys in practice, prior to designing conservation works, and of continuing research into erosion survey techniques is that they perform the following tasks more effectively than any other method. First, they provide a focus for integrating research on erosion carried out by geomorphologists, hydrologists, pedologists and engineers. Second, they draw attention to the most relevant factors controlling erosion at different scales of study and, therefore, to those to which conservation measures need be directed. Third, they help identify the most important processes at work. This is vital for deciding whether anti-erosion measures should be concentrated on surface or subsurface erosion, on detachment or transport of soil particles, and on promoting infiltration of excess water or its removal over the land surface in waterways. Fourth, they point to the problem of selecting an appropriate figure as a target to which to reduce the erosion rate. There can be no universally applicable figure. It must be varied according to the rate of natural erosion, which in turn is related to the relief of the land and the frequency of moderate and high magnitude wind and rain storms, and the size of the area being considered. Fifth, they provide a system of mapping the data required to develop predictive models.

Although the emphasis of this book has been placed, intentionally, on soil erosion, its philosophy has been to encourage better conservation practice and more acceptable conservation design. Improvements to conservation techniques are most likely to come from the research of agricultural engineers. Indeed, soil conservation is traditionally a preserve of agricultural engineering. It is hoped that this book has shown the valuable role, in supplementing and making more effective the work of the engineer, that the applied physical geographer can play.

REFERENCES

Ahnert, F. (1976) 'Brief description of a comprehensive three-dimensional process-response model of landform development', *Z. f. Geomorph. Suppl.* **25,** 29–49.

André, J. E. and Anderson, H. W. (1961) 'Variation of soil erodibility with geology, geographic zone, elevation and vegetation type in northern California wildlands', *J. geophys. Res.* **66,** 3351–8.

Ateshian, J. K. H. (1974) 'Estimation of rainfall erosion index', *J. Irrigation and Drainage Div. ASCE,* **100,** 293–307.

Bagnold, R. A. (1937) 'The transport of sand by wind', *Geogr. J.* **89,** 409–38.

Bayfield, N. G. (1973) 'Use and deterioration of some Scottish hill paths', *J. appl. Ecol.* **10,** 635–44.

Bergsma, E. (1974) 'Soil erosion sequences on aerial photographs', *ITC Journal* **1974–3,** 342–76.

Berry, L. (1970) 'Some erosional features due to piping and subsurface wash with special reference to the Sudan', *Geografiska Ann.* **52–A,** 113–19.

Berry, L. and Ruxton, B. P. (1960) 'The evolution of Hong Kong harbour basin', *Z. f. Geomorph.* **4,** 97–115.

Berry, M. J. (1956) 'Erosion control on Bukit Bakar, Kelantan', *Malay. Forester* **19,** 3–11.

Bie, S. W. and Beckett, P. H. T. (1971) 'Quality control in soil survey. II. The cost of soil survey', *J. Soil Sci.* **22,** 453–65.

Blong, R. J. (1970) 'The development of discontinuous gullies in a pumice catchment', *Am. J. Sci.* **268,** 369–83.

Borst, H. L. and Woodburn, R. (1942) 'The effect of mulching and methods of cultivation on runoff and erosion from Muskingum silt loam', *Agr. Engr.* **23,** 19–22.

Borsy, Z. (1972) 'Studies on wind erosion in the wind-blown sand areas of Hungary', *Acta Geographica Debrecina* **10,** 123–32.

Bouyoucos, G. J. (1935) 'The clay ratio as a criterion of susceptibility of soils to erosion', *J. Am. Soc. Agron.* **27,** 738–41.

Brice, J. C. (1966) 'Erosion and deposition in the loess-mantled Great Plains, Medicine Creek drainage basin, Nebraska', *U.S. geol. Surv. Prof. Paper* **352–H,** 255–339.

Bryan, R. B. (1968) 'The development, use and efficiency of indices of soil erodibility', *Geoderma* **2,** 5–26.

Bryan, R. B. (1969) 'The relative erodibility of soils developed in the Peak District of Derbyshire', *Geografiska Ann.* **51–A,** 145–59.

Buckham, A. F. and Cockfield, W. E. (1950) 'Gullies formed by sinking of the ground', *Am. J. Sci.* **248,** 137–41.

Burgess, P. F. (1971) 'The effect of logging on hill dipterocarp forest', *Malay. Nat. J.* **24,** 231–7.

Burrough, P. A., Beckett, P. H. T. and Jarvis, M. G. (1971) 'The relation between cost and utility in soil survey (I–III)', *J. Soil Sci.* **22,** 359–94.

Burykin, A. M. (1957) 'O vnutripochvennom stoke v gornykh usloviyakh vlazhnykh subtropikov', *Pochvovedenie* **12,** Engl. transl. in *Water regime and erosion,* Israel Program for Scientific Translation, Nat. Sci. Foundation and Dept. Agric., Washington, D.C.

Carson, M. A. and Kirkby, M. J. (1972) *Hillslope form and process,* Cambridge University Press.

Carter, C. E., Greer, J. D., Braud, H. J. and Floyd, J. M. (1974) 'Raindrop characteristics in south central United States', *Trans. Am. Soc. Agr. Engr.* **17,** 1033–7.

Chepil, W. S. (1945) 'Dynamics of wind erosion. III. Transport capacity of the wind', *Soil Sci.* **60,** 475–80.

Chepil, W. S. (1950) 'Properties of soil which influence wind erosion. II. Dry aggregate structure as an index of erodibility', *Soil Sci.* **69**, 403–14.

Chepil, W. S. and Woodruff, N. P. (1963) 'The physics of wind erosion and its control', *Advances in Agronomy* **15**, 211–302.

Chorley, R. J. (1959) 'The geomorphic significance of some Oxford soils', *Am. J. Sci.* **257**, 503–15.

Cooke, R. U. and Doornkamp, J. C. (1974) *Geomorphology in environmental management,* Oxford University Press, London.

David, W. P. and Beer, C. E. (1975) 'Simulation of soil erosion – Part 1. Development of a mathematical erosion model', *Trans. Am. Soc. Agr. Engr.* **18**, 126–9, 133.

Demek, J. (1971) *Manual of detailed geomorphological mapping,* Czechoslovak Academy of Science, Institute of Geography, Brno.

De Ploey, J. (1974) 'Mechanical properties of hillslopes and their relation to gullying in central semi-arid Tunisia', *Z. f. Geomorph. Suppl.* **21**, 177–90.

De Ploey, J. (1977) 'Some experimental data on slopewash and wind action with reference to Quaternary morphogenesis in Belgium', *Earth Surface Processes* **2**, 101–15.

De Ploey, J. and Gabriels, D. (in press) 'Methods of measuring soil loss', Chapter 3 in Kirkby, M. J. and Morgan, R. P. C. (eds), *Soil erosion,* to be published by John Wiley, London.

De Ploey, J., Savat, J. and Moeyersons, J. (1976) 'The differential impact of some soil factors on flow, runoff creep and rainwash', *Earth Surface Processes* **1**, 151–61.

Douglas, I. (1967a) 'Natural and man-made erosion the humid tropics of Australia, Malaysia and Singapore', *Int. Assoc. scient. Hydrol. Pub.* **75**, 17–30.

Douglas, I. (1976b) 'Man, vegetation and sediment yield of rivers', *Nature* **215**, 925–8.

Douglas, I. (1968a) 'Sediment sources and causes in the humid tropics of northeast Queensland', in Harvey, A. M. (ed.), *Geomorphology in a tropical environment,* British Geomorphological Research Group, *Occasional Paper* **5**, 27–39.

Douglas, I. (1968b) 'Erosion in the Sungai Gombak catchment, Selangor, Malaysia', *J. trop. Geog.* **26**, 1–16.

Douglas, I. (1970) 'Measurements of river erosion in West Malaysia', *Malay. Nat. J.* **23**, 78–83.

Douglas, I. (1972) *The environment game,* Inaugural lecture, University of New England, Armidale.

Douglas, I. (1976) 'Erosion rates and climate: geomorphological implications', in Derbyshire, E. (ed.), *Geomorphology and climate,* John Wiley, London, 269–87.

Drainage and Irrigation Department (1970) *Rainfall records for West Malaysia, 1959—1965,* Kuala Lumpur.

D'Souza, V. P. C. and Morgan, R. P. C. (1976) 'A laboratory study of the effect of slope steepness and curvature on soil erosion', *J. agric. Engng. Res.* **21**, 21–31.

Dunne, T. and Black, R. D. (1970) 'Partial area contributions to storm runoff in a small New England watershed', *Water Resources Research* **6**, 1296–311.

Edgar, A. T. (1958) *Manual of rubber planting,* Incorporated Society of Planters, Kuala Lumpur.

Ellison, W. D. (1947) 'Soil erosion studies', *Agr. Engr.* **28**, 145–6, 197–201, 245–8, 297–300, 349–51.

Elwell, H. A. and Stocking, M. A. (1976) 'Vegetal cover to estimate soil erosion hazard in Rhodesia', *Geoderma* **15**, 61–70.

Emmett, W. W. (1970) 'The hydraulics of overland flow on hillslopes', *U.S. geol. Surv. Prof. Paper* **662–A.**

Evans, R. (in press) 'Mechanics of water erosion', Chapter 4 in Kirkby, M. J. and Morgan, R. P. C. (eds), *Soil erosion,* to be published by John Wiley, London.

Evans, R. and Morgan, R. P. C. (1974) 'Water erosion of arable land', *Area* **6**, 221–5.

Eyles, R. J. (1966) 'Stream representation on Malayan maps', *J. trop. Geog.* **22**, 1–9.

Eyles, R. J. (1967) 'Laterite at Kerdau, Pahang, Malaya', *J. trop. Geog.* **25**, 18–23.

Eyles, R. J. (1968a) 'Stream net ratios in West Malaysia', *Bull. geol. Soc. Am.* **79**, 701–12.

Eyles, R. J. (1968b) *A morphometric analysis of West Malaysia,* unpublished PhD thesis, Univeristy of Malaya.

Eyles, R. J. (1968c) 'Morphometric explanation: a case study', *Geographica (University of Malaya)* **4**, 17–23.

Eyles, R. J. (1970) 'Physiographic implications of laterite', *Bull. geol. Soc. Malay.* **3**, 1–7.

F.A.O. (1965) *Soil erosion by water,* Rome.

Farmer, E. E. (1973) 'Relative detachability of soil particles by simulated rainfall', *Soil Sci. Soc. Am. Proc.* **37**, 629–33.

Fermor, L. L. (1939) *Report upon the mining industry of Malaya,* Govt. Printer, Kuala Lumpur.

Foster, G. R. and Meyer, L. D. (1975) 'Mathematical simulation of upland erosion by fundamental erosion mechanics', in *Present and prospective technology for predicting sediment yields and sources,* U.S. Dept. Agr., Agr. Res. Service, Pub. **ARS–S–40,** 190–207.

Foster, G. R., Meyer, L. D. and Onstad, C. A. (1973) 'Erosion equations derived from modelling principles', *ASAE Paper No. 73—2550,* Am. Soc. Agr. Engnr. St. Joseph, Mich.

Fournier, F. (1960) *Climat et érosion: la relation entre l'érosion du sol par l'eau et les précipitations atmosphériques,* Presses Universitaires de France, Paris.

Fournier, F. (1972) *Soil conservation,* Nature and Environment Series, Council of Europe.

Gabriels, D., Pauwels, J. M. and De Boodt, M. (1975) 'The slope gradient as it affects the amount and size distribution of soil loss material from runoff on silt loam aggregates', *Med. Fac. Landbouww. Rijksuniv. Gent* **40,** 1333–8.

Gabriels, D., Pauwels, J. M. and De Boodt, M. (1977) 'A quantitative rill erosion study on a loamy sand in the hilly region of Flanders', *Earth Surface Processes* **2,** 257–9.

Gerlach, T. (1966) 'Współczesny rozwój stoków w dorzeczu górnego Grajcarka (Beskid Wysoki – Karpaty Zachodnie)', *Prace Georgr. IG PAN* **52** (with French summary).

Gerlach, T. (1967) 'Evolutions actuelles des versants dans les Carpathes, d'après l'exemple d'observations fixes', in Macar, P. (ed.), *L'évolution des versants,* University of Liège, 129–38.

Gerlach, T. and Niemirowski, M. (1968) 'Charakterystyka geomorfologiczna dolin Jaszcze i Jamne', *Zak. Ochr. Przyrody Polsk. Ak. Nauk, Studia Naturae, Seria A* **2,** 11–22.

González, M. H. (1971) '¿Que es condición de pastizal?', *Boletín Pastizales* **2,** 8–10.

Gregory, K. J. and Walling, D. E. (1973) *Drainage basin form and process,* Edward Arnold, London.

Hack, J. T. and Goodlett, J. C. (1960) 'Geomorphology and forest ecology of a mountain region in the central Appalachians', *U.S. geol Surv. Prof. Paper* **347.**

Hall, M. J. (1970) 'A critique of methods of simulating rainfall', *Water Resources Research* **6,** 1104–14.

Hartley, C. W. S. (1949) 'Soil erosion in Malaya', *Corona,* **1,** 25–7.

Heede, B. H. (1971) 'Characteristics and processes of soil piping in gullies', *USDA Forest Service, Research Paper* **RM–68,** Rocky Mountain Forest and Range Experiment Station, Fort Collins, Colo.

Heede, B. H. (1975) 'Stages of development of gullies in the west', in *Present and prospective technology for predicting sediment yields and sources,* U.S. Dept. Agr., Agr. Res. Service, Pub. **ARS–S–40,** 155–61.

Heede, B. H. (1976) 'Gully development and control: the status of our knowledge', *USDA Forest Service, Research Paper* **RM–169,** Rocky Mountain Forest and Range Experiment Station, Fort Collins, Colo.

Heusch, B. (1970) L'érosion du Pré-Rif. Une étude quantitative de l'érosion hydraulique dans les collines marneuses du Pré-Rif occidental', *Ann. Rech. fores. Maroc* **12,** 9–176.

Higginson, F. R. (1973) 'Soil erosion of land systems within the Hunter Valley', *J. Soil Conserv. Serv. New South Wales* **29,** 103–10.

Hills, R. C. (1970) 'The determination of the infiltration capacity of field soils using the cylinder infiltrometer', *British Geomorphological Research Group, Technical Bulletin* **No. 5.**

Hjulström, F. (1935) 'Studies of the morphological activity of rivers as illustrated by the River Fyries', *Bull. geol. Inst. Univ. Uppsala* **25,** 221–527.

Horton, R. E. (1945) 'Erosional development of streams and their drainage basins: a hydrophysical approach to quantitative morphology', *Bull. geol. Soc. Am.* **56,** 275–370.

Horváth, V. and Erödi, B. (1962) 'Determination of natural slope category limits by functional identity of erosion intensity', *Int. Assoc. scient. Hydrol. Pub.* **59,** 131–43.

Hudson, N. W. (1957) 'The design of field experiments on soil erosion', *J. agric. Engng. Res.* **2,** 56–65.

Hudson, N. W. (1961) 'An introduction to the mechanics of soil erosion under conditions of sub-tropical rainfall', *Proceedings and Transactions, Rhod. scient. Assoc.* **49,** 15–25.

Hudson, N. W. (1963) 'Raindrop size distribution in high intensity storms', *Rhod. J. agric. Res.* **1,** 6–11.

Hudson, N. W. (1965) *The influence of rainfall on the mechanics of soil erosion with particular reference to Southern Rhodesia,* unpublished MSc thesis, University of Cape Town.

Hudson, N. W. (1971) *Soil conservation,* Batsford, London.

Hudson, N. W. and Jackson, D. C. (1959) 'Results achieved in the measurement of erosion and runoff in Southern Rhodesia', *Proceedings, Third Inter-African Soils Conference, Dalaba,* 575–83.

Iana, S. (1972) 'Considérations sur la protection des versants en Dobroudgea', *Acta Geographica Debrecina* **10,** 51–5.

Imeson, A. C. (1971) 'Heather burning and soil erosion on the North Yorkshire Moors', *J. appl. Ecol.* **8**, 537–42.

Jones, R. G. B. and Keech, M. A. (1966) 'Identifying and assessing problem areas in soil erosion surveys using aerial photographs', *Photogrammetric Record* **5/27**, 189–97.

Josué Martínez, T. G. and González, M. H. (1971) 'Influencia de la condición de pastizal en la infiltración de agua en el suelo', *Boletín Pastizales* **2 (2)**, 2–5.

Jovanović, S. and Vukčević, M. (1958) 'Suspended sediment regimen on some watercourses in Yugoslavia and analysis of erosion processes', *Int. Assoc. scient. Hydrol. Pub.* **43**, 337–59.

Keech, M. A. (1969) 'Mondaro Tribal Trust Land. Determination of trend using air photo analysis', *Rhod. agric. J.* **66**, 3–10.

Kellman, M. C. (1969) 'Some environmental components of shifting cultivation in upland Mindanao', *J. trop. Geog.* **28**, 40–56.

Kirkby, M. J. (1969*a*) 'Infiltration, throughflow and overland flow', in Chorley, R. J. (ed.), *Water, earth and man*, Methuen, London, 215–27.

Kirkby, M. J. (1969*b*) 'Erosion by water on hillslopes' in Chorley, R. J. (ed.), *Water, earth and man*, Methuen, London, 229–38.

Kirkby, M. J. (1971) 'Hillslope process-response models based on the continuity equation', in Brunsden, D. (ed.), *Slopes: form and process*, Inst. Br. Geogr. Special Pub. **3**, 15–30.

Kirkby, M. J. (1976) 'Hydrological slope models: the influence of climate', in Derbyshire, E. (ed.), *Geomorphology and climate*, John Wiley, London, 247–67.

Klingebiel, A. A. and Montgomery, P. H. (1966) 'Land capability classification', *USDA Soil Conserv. Serv. Agr. Handbook* **210**.

Kohnke, H. and Bertrand, A. R. (1959) *Soil conservation*, McGraw-Hill, New York.

Laflen, J. M., Johnson, H. P. and Reeve, R. C. (1972) 'Soil loss from tile-outlet terraces', *J. Soil and Wat. Conserv.* **27**, 74–7.

Lake, H. M. (1894) 'Johore', *Geogr. J.* **3**, 281–302.

Lal, R. (1976) 'Soil erosion problems on an alfisol in western Nigeria and their control', *IITA Monograph* **1**.

Langbein, W. B. and Schumm, S. A. (1958) 'Yield of sediment in relation to mean annual precipitation', *Trans. Am. geophys. Un.* **39**, 1076–84.

Lattanzi, A. R., Meyer, L. D. and Baumgardner, M. F. (1974) 'Influence of mulch rate and slope steepness on interrill erosion', *Soil Sci. Soc. Am. Proc.* **38**, 946–50.

Laws, J. O. and Parsons, D. A. (1943) 'The relationship of raindrop size to intensity', *Trans. Am. geophys. Un.* **24**, 452–60.

Leaf, C. F. (1970) 'Sediment yields from central Colorado snow zone', *J. Hydraulics Division, ASCE* **96**, 87–93.

Leopold, L. B., Wolman, M. G. and Miller, J. P. (1964) *Fluvial processes in geomorphology*, Freeman, San Francisco.

Low, F. K. (1967) 'Estimating potential erosion in developing countries', *J. Soil and Wat. Conserv.* **22**, 147–8.

Low, K. S. and Goh, K. C. (1972) 'The water balance of five catchments in Selangor, West Malaysia', *J. trop. Geog.* **35**, 60–6.

Lundgren, L. and Rapp, A. (1974) 'A complex landslide with destructive effects on the water supply of Morogoro town, Tanzania', *Geografiska Ann.* **56–A**, 251–60.

McCormack, R. J. (1971) 'The Canada Land Use Inventory: a basis for landuse planning', *J. Soil and Wat. Conserv.* **26**, 141–6.

Maheswari, V. (1970) 'Deviation of the Sungai Mulia', *Geographica (University of Malaya)* **6**, 123–9.

Mason, B. J. and Andrews, J. B. (1960) 'Drop size distributions from various types of rain', *Quart. J. Roy. Met. Soc.* **86**, 346–53.

Meyer, L. D. (1965) 'Mathematical relationships governing soil erosion by water', *J. Soil and Wat. Conserv.* **20**, 149–50.

Meyer, L. D. and Wischmeier, W. H. (1969) 'Mathematical simulation of the process of soil erosion by water', *Trans. Am. Soc. Agr. Engnr.* **12**, 754–8, 762.

Middleton, H. E. (1930) 'Properties of soils which influence soil erosion', *U.S. Dept. Agr. Technical Bulletin* **178**.

Mikhailov, T. (1972) 'Certaines particularités des processus d'érosion contemporains en Bulgarie', *Acta Geographica Debrecina* **10**, 41–50.

Moeyersons, J. and De Ploey, J. (1976) 'Quantitative data on splash erosion, simulated on unvegetated slopes', *Z. f. Geomorph. Suppl.* **25**, 120–31.

Morgan, R. P. C. (1971) Rainfall of West Malaysia: a preliminary regionalization using principal components analysis', *Area* **3**, 222–7.

Morgan, R. P. C. (1972) 'Observations on factors affecting the behaviour of a first-order stream', *Trans. Inst. Br. Geogr.* **56**, 171–85.

Morgan, R. P. C. (1973) 'Soil–slope relationships in the lowlands of Selangor and Negri Sembilan, West Malaysia', *Z. f. Geomorph.* **17**, 139–55.

Morgan, R. P. C. (1974) 'Estimating regional variations in soil erosion hazard in Peninsular Malaysia', *Malay. Nat. J.* **28**, 94–106.

Morgan, R. P. C. (1976) 'The role of climate in the denudation system: a case study from West Malaysia', in Derbyshire, E. (ed.), *Climate and geomorphology,* John Wiley, London, 317–43.

Morgan, R. P. C. (1977) 'Soil erosion in the United Kingdom: field studies in the Silsoe area, 1973–75', *Nat. Coll. Agr. Engng. Silsoe, Occasional Paper* **5**.

Morgan, R. P. C. (1978) 'Field studies of rainsplash erosion', *Earth Surface Processes,* **3**, 259–9.

Morgan, R. P. C. (in press) 'Implications', Chapter 8 in Kirkby, M. J. and Morgan, R. P. C. (eds), *Soil erosion,* to be published by John Wiley, London.

Morgan, R. P. C. and Keech, M. A. (1976) 'Aspects of scale in the application of remote sensing to the evaluation of soil erosion risk', in Collins, W. G. and van Genderen, J. L. (eds), *Landuse studies by remote sensing,* Remote Sensing Society, 40–53.

Morin, J., Goldberg, D. and Seginer, I. (1967) 'A rainfall simulator with a rotating disc', *Trans. Am. Soc. Agr. Engnr.* **10**, 74–7, 79.

Mosley, M. P. (1974) 'Experimental study of rill erosion', *Trans. Am. Soc. Agr. Engnr.* **17**, 909–13, 916.

Musgrave, G. W. (1947) 'The quantitative evaluation of factors in water erosion: a first approximation', *J. Soil and Wat. Conserv.* **2**, 133–8.

Mutchler, C. K. and Young, R. A. (1975) 'Soil detachment by raindrops', in *Present and prospective technology for predicting sediment yields and sources,* U.S. Dept. Agr., Agr. Res. Service, Pub. **ARS–S–40**, 113–17.

Nassif, S. H. and Wilson, E. M. (1975) 'The influence of slope and rain intensity on runoff and infiltration', *Hydrol. Science Bull.* **20**, 539–53.

Negev, M. (1967) 'A sediment model on a digital computer', *Stanford Univ. Dept. Civil Engng. Technical Report* **76**.

Ng, S. K. (1969) 'Soil resources in Malaya' in Stone, B. C. (ed.), *Natural resources in Malaysia and Singapore,* Kuala Lumpur, 141–51.

Nossin, J. J. (1964) 'Geomorphology of the surroundings of Kuantan (Eastern Malaya)', *Geologie en Mijnbouw* **43**, 157–82.

Olofin, E. A. (1972) *Landform analysis of the Ulu Langat District, Selangor: a case study in map compilation for planning purposes,* unpublished MA thesis, University of Malaya.

Olofin, E. A. (1974) 'Classification of slope angles for land planning purposes', *J. trop. Geog.* **39**, 72–7.

Ologe, K. O. (1972) 'Gullies in the Zaria area. A preliminary study of headscarp recession', *Savanna* **1**, 55–66.

Palmer, R. S. (1964) 'The influence of a thin water layer on waterdrop impact forces', *Int. Assoc. scient. Hydrol. Pub.* **65**, 141–8.

Panton, W. P. (1969) 'Land capability classification programme for West Malaysia', in Stone, B. C. (ed.), *Natural resources in Malaysia and Singapore,* Kuala Lumpur, 152–60.

Parsons, D. A., Apmann, R. P. and Decker, G. H. (1964) 'The determination of sediment yields from flood water sampling', *Int. Assoc. scient. Hydrol. Pub.* **65**, 7–15.

Pearce, A. J. (1976) 'Magnitude and frequency of erosion by Hortonian overland flow', *J. Geology,* **84**, 65–80.

Quinn, N. W., Morgan, R. P. C. and Smith, A. J. (in preparation).

Rao, D. P. (1975) 'Applied geomorphological mapping for erosion surveys: the example of the Oliva basin, Calabria', *ITC Journal,* **1975–3**, 341–50.

Rapp, A., Axelsson, V., Berry, L. and Murray-Rust, D. H. (1972) 'Soil erosion and sediment transport in the Morogoro river catchment, Tanzania', *Geografiska Ann.* **54–A**, 125–55.

Reynolds, S. G. (1975) 'Soil property variability in slope studies: suggested sampling schemes and typical required sample sizes', *Z. f. Geomorph.* **19**, 191–208.

Rich, L. R. (1972) 'Managing a ponderosa pine forest to increase water yield', *Water Resources Research* **8**, 422–8.

Richter, G. and Negendank, J. F. W. (1977) 'Soil erosion processes and their measurement in the German area of the Moselle river', *Earth Surface Processes* **2**, 261–78.

Roose, E. J. (1966) *Etude de la méthode des bandes d'arrêt pour la conservation de l'eau et des sols,* Cyclo. ORSTOM, Adiopodoumé, Ivory Coast.

Roose, E. J. (1967) 'Dix années de mesure de l'érosion et du ruissellement au Sénégal', *L'Agron. Trop.* **22**, 123–52.

Roose, E. J. (1970) 'Importance relative de l'érosion, du drainage oblique et vertical dans la pédogenèse actuelle d'un sol ferrallitique de moyenne Côte d'Ivoire', *Cah. ORSTOM, sér. Pédol.* **8**, 469–82.

Roose, E. J. (1971) *Influence des modifications du millieu naturel sur l'érosion: le bilan hydrique et chimique suite à la mise en culture sous climat tropical,* Cyclo. ORSTOM, Adiopodoume, Ivory Coast.

Roose, E. J. (1975) *Erosion et ruissellement en Afrique de l'ouest: vingt années de mesures en petites parcelles expérimentales,* Cyclo. ORSTOM, Abidjan, Ivory Coast.

Rubey, W. W. (1952) 'Geology and mineral resources of the Hardin and Brussels Quadrangles, Illinois', *U.S. geol. Surv. Prof. Paper,* **218**.

Sánchez Muñoz, A. and Valdés Reyna, J. (1975) 'Infiltración de agua en dos tipos vegetativos relacionando suelo-vegetación, *Boletín Pastizales* **6 (5)**, 2–6.

Savat, J. (1977) 'The hydraulics of sheet flow on a smooth surface and the effect of simulated rainfall', *Earth Surface Processes* **2**, 125–40.

Schwab, G. O., Frevert, R. K., Edminster, T. W. and Barnes, K. K. (1966) *Soil and water conservation engineering,* John Wiley, New York.

Shallow, P. G. D. (1956) 'River flow in the Cameron Highlands', *Central Electricity Board, Hydro-electric Technical Memorandum* **3**, Kuala Lumpur.

Sharpe, C. F. S. (1938) *Landslides and related phenomena,* Columbia University Press, New York.

Sheng, T. C. (1972*a*) 'A treatment-oriented land capability classification scheme for hilly marginal lands in the humid tropics', *J. scient. Research Council, Jamaica* **3**, 93–112.

Sheng, T. C. (1972*b*) 'Bench terracing', *J. scient. Research Council, Jamaica* **3**, 113–27.

Shiyatyy, Ye. I., Lavrovskiy, A. B. and Khmolenko, M. I. (1972) 'Effect of texture on the cohesion and wind resistance of soil clods', *Soviet Soil Science* **4**, 105–12.

Skidmore, E. L. and Woodruff, N. P. (1968) 'Wind erosion forces in the United States and their use in predicting soil loss', *USDA Agr. Res. Serv. Agr. Handbook* **346**.

Smalley, I. J. (1970) 'Cohesion of soil particles and the intrinsic resistance of simple soil systems to wind erosion', *J. Soil Sci.* **21**, 154–61.

Smith, D. D. (1958) 'Factors affecting rainfall erosion and their evaluation', *Int. Assoc. scient. Hydrol. Pub.* **43**, 97–107.

Soons, J. M. and Rainer, J. N. (1968) 'Micro-climate and erosion processes in the Southern Alps, New Zealand', *Geografiska Ann.* **50–A**, 1–15.

Soper, J. R. P. (1938) 'Soil erosion on Penang Hill', *Malay. Agr. J.* **26**, 407–13.

Speer, W. S. (1963) 'Report to the Government of Malaysia on soil and water conservation', *FAO Report* **1788**, Rome.

Stallings, J. H. (1957) *Soil conservation,* Prentice-Hall, Englewood Cliffs, N.J.

Starkel, L. (1972) 'The role of catastrophic rainfall in the shaping of the relief of the Lower Himalaya (Darjeeling Hills)', *Geogr. Polonica* **21**, 103–47.

Starkel, L. (1976) 'The role of extreme (catastrophic) meteorological events in contemporary evolution of slopes', in Derbyshire, E. (ed.), *Geomorphology and climate,* John Wiley, London, 203–46.

Stocking, M. A. (1972) 'Relief analysis and soil erosion in Rhodesia using multivariate techniques', *Z. f. Geomorph.* **16**, 432–43.

Stocking, M. A. and Elwell, H. A. (1973*a*) 'Prediction of subtropical storm soil losses from field plot studies', *Agric. Met.* **12**, 193–201.

Stocking, M. A. and Elwell, H. A. (1973*b*) 'Soil erosion hazard in Rhodesia', *Rhod. agric. J.* **70**, 93–101.

Stocking, M. A. and Elwell, H. A. (1976) 'Rainfall erosivity over Rhodesia', *Trans. Inst. Br. Geogr. New Series* **1**, 231–45.

Swan. S. B. St. C. (1970) 'Piedmont slope studies in a humid tropical region, Johor, southern Malaya', *Z. f. Georph. Suppl.* **10**, 30–9.

Temple, P. H. (1972*a*) 'Measurements of runoff and soil erosion at an erosion plot scale with particular reference to Tanzania', *Geografiska Ann.* **54–A**, 203–20.

Temple, P. H. (1972*b*) 'Soil and water conservation policies in the Uluguru Mountains, Tanzania', *Geografiska Ann.* **54–A**, 110–23.

Temple, P. H. and Murray-Rust, D. H. (1972) 'Sheet wash measurements on erosion plots at Mfumbwe, eastern Uluguru Mountains, Tanzania', *Geografiska Ann.* **54–A**, 195–202.

Temple, P. H. and Rapp, A. (1972) 'Landslides in the Mgeta area, western Uluguru Mountains, Tanzania', *Geografiska Ann.* **54–A**, 157–93.

110 Soil erosion

Thornes, J. B. (1976) 'Semi-arid erosion systems: case studies from Spain', *London School of Economics, Geogr. Papers* **7**.
Tricart, J. (1961) 'Mécanismes normaux et phénomènes catastrophiques dans l'évolution des versants du bassin du Guil (Hautes-Alpes, France)', *Z. f. Geomorph.* **5**, 277–301.
Tricart, J. (1972) *Landforms of the humid tropics, forests and savannas,* transl. de Jonge, C. J. K. Longman, London.
Tuckfield, C. G. (1964) 'Gully erosion in the New Forest, Hampshire', *Am. J. Sci.* **262**, 759–807.
van Genderen, J. L. (1970) *The morphodynamics of the Crati River basin, Calabria,* ITC Publications, Series B, No. 56, Delft.
van Wambeke, A. R. (1962) 'Criteria for classifying tropical soils by age', *J. Soil Sci.* **13**, 124–32.
Verstappen, H. Th. and van Zuidam, R. A. (1968) *ITC system of geomorphological survey,* ITC, Delft.
Vink, A. P. A. (1968) 'Aerial photographs and the soil sciences', in *Aerial surveys and integrated studies,* UNESCO, Paris, 127–31.
Vink, A. P. A. (1975) *Landuse in advancing agriculture,* Springer-Verlag, Berlin.
Vittorini, S. (1972) 'The effects of soil erosion in an experimental station in the Pliocene clay of the Val d'Era (Tuscany) and its influence on the evolution of the slopes', *Acta Geographica Debrecina* **10**, 71–81.
Voznesensky, A. S. and Artsruui, A. B. (1940) 'A laboratory method for determining the anti-erosion resistance of soils' (in Russian), *Tiflis* 18–33, abstract in *Soils and Fertilizers* **10**, 289.
Vuillaume, G. (1969) 'Analyse quantitative du rôle du milieu physico-climatique sur le ruissellement et l'érosion à l'issue de bassins de quelques hectares en zone sahélienne (Bassin du Kountkouzout, Niger)', *Cah. ORSTOM, sér. Hydrol.* **6**, 87–132.
Walling, D. E. (1974) 'Suspended sediment and solute yields from a small catchment prior to urbanisation', in Gregory, K. J. and Walling, D. E. (eds), *Fluvial processes in instrumented watersheds,* Inst. Br. Geogr. Special Pub. **6**, 169–92.
Whitmore, T. C. and Burnham, C. P. (1969) 'The altitudinal sequence of forests and soils on granite near Kuala Lumpur', *Malay. Nat. J.* **22**, 99–118.
Williams, A. R. and Morgan, R. P. C. (1976) 'Geomorphological mapping applied to soil erosion evaluation', *J. Soil and Wat. Conserv.* **31**, 164–8.
Williams, C. N. and Joseph, K. T. (1970) *Climate, soil and crop production in the humid tropics,* Oxford University Press, Kuala Lumpur.
Wischmeier, W. H. and Mannering, J. V. (1969) 'Relation of soil properties to its erodibility', *Soil Sci. Soc. Am. Proc.* **23**, 131–7.
Wischmeier, W. H. and Smith, D. D. (1958) 'Rainfall energy and its relationship to soil loss', *Trans. Am. geophys. Un.* **39**, 285–91.
Wischmeier, W. H. and Smith, D. D. (1962) 'Soil loss estimation as a tool in soil and water management planning', *Int. Assoc. scient. Hydrol. Pub.* **59**, 148–59.
Wischmeier, W. H. and Smith, D. D. (1965) 'Predicting rainfall erosion losses from cropland east of the Rocky Mountains', *USDA, Agr. Res. Serv. Agr. Handbook* **282**.
Wischmeier, W. H., Johnson, C. B. and Cross, B. V. (1971) 'A soil erodibility nomograph for farmland and construction sites', *J. Soil and Wat. Conserv.* **26**, 189–93.
Wolman, M. G. (1967) 'A cycle of sedimentation and erosion in urban river channels', *Geografiska Ann.* **49–A**, 385–95.
Woodburn, R. and Kozachyn, J. (1956) 'A study of relative erodibility of a group of Mississippi gully soils', *Trans. Am. geophys. Un.* **37**, 749–53.
Yaïr, A. (1972) 'Observations sur les effets d'un ruissellement dirigé selon la pente des interfluves dans une région semi-aride d'Israël', *Rev. Géogr. phys. Géol. dyn.* **14**, 537–48.
Yariv, S. (1976) 'Comments on the mechanism of soil detachment by rainfall', *Geoderma* **15**, 393–9.
Young, A. (1969) 'Present rate of land erosion', *Nature* **224**, 851–2.
Young, A. (1972) *Slopes,* Oliver and Boyd, Edinburgh.
Young, A. (1973) 'Soil survey procedure in land development planning', *Geogr. J.* **139**, 53–64.
Young, A. (1976) *Tropical soils and soil survey,* Cambridge University Press.
Zaborski, B. (1972) 'On the origin of gullies in loess', *Acta Geographica Debrecina* **10**, 109–111.
Zaruba, Q. and Mencl, V. (1969) *Landslides and their control,* Elsevier, New York.
Zingg, A. W. (1940) 'Degree and length of land slope as it affects soil loss in runoff', *Agr. Engr.* **21**, 59–64.

INDEX

acceptable rates of erosion, 69, 72, 79, 88, 89–90, 95, 103
aerial photography, 36, 39, 72, 80–1, 82–3, 85, 97
aggregate stability, 21–3, 57, 61, 86
agronomic conservation measures, 57–61, 67, 68, 69, 88

bank erosion, 11, 71, 81, 83, 91

clay minerals, 21
climate, as related to erosion rates, 3, 50
continuity equation, 55
contour bunds, 40, 63
contour ploughing, 62
contour strip cropping, 51
cover crops, 59, 90, 95, 99, 100
crusting, of soil surface, 6, 9, 22, 39

detachment, 5, 6–7, 15
    effects on conservation design, 57, 69, 103
    effects on erodibility, 21, 88
    inclusion in erosion models, 55
detachment capacity, 9, 14
detachment-limited condition, 5
diversion channels, 66–7, 101
divisors, 44
drainage, 62
drainage density, 3, 49, 72–3, 76, 89
drainage texture, 72–3, 76

effects of erosion, 2
energy, kinetic, 5–6
    dissipation by plant cover, 25
    estimation of, 18–20, 27, 72–3
    in rainfall simulation, 45
energy, potential, 5
erodibility, 21–4, 39, 88, 97
    as factor affecting erosion, 1, 15

as factor in erosion hazard assessment, 29, 99
    effects on conservation design, 61
    in Universal Soil-Loss Equation, 51
erosion
    hazard assessment, 26–40, 72
    plots, 42–4
    risk, 70, 72; see also erosion hazard assessment and erosion surveys
    surveys, 26, 36–40, 70, 79, 80–92, 93–102, 103
erosivity, 15–18
    as factor affecting erosion, 1
    as factor in erosion hazard assessment, 28, 30, 37, 77, 99
    effects on conservation design, 100
    indices for rainfall, 18–20, 26–7, 73–6
    indices for wind, 20–1
    in Universal Soil-Loss Equation, 51, 54

feasibility studies, 95–7, 102
fertilizers, 2, 10, 61
field
    experiments, 42–5
    surveys, 39, 80, 83–8, 97
fire, 11, 58
floods, 2, 18, 59, 78–9
forests
    as conservation measure, 78, 100
    effects on erosion rates, 2, 25, 72, 89–90, 95
    see also timber exploitation
Fournier's prediction equation, 27, 50, 79, 88
frequency and magnitude of erosion, 2–3, 17–18, 69, 76, 98
Froude number, 7

gabions, 62
Gerlach gutters, 44–5
geomorphological mapping, 37
gleying, 60
grass waterways, 40, 66–7, 101

gully
 density, 37, 39, 40, 72, 76
 erosion, 2, 5, 55, 58, 59, 71
  as factor in erosion hazard assessment,
   36–9, 76–7, 79, 80–1, 83, 89, 97, 99
  control of, 67, 68, 90, 100
  effects of antecedent events, 17
  effects on conservation design, 90
  frequency of, 18
  inclusion in erosion models, 55
  processes of, 10, 11, 13, 98
 networks, 11
 reclamation, *see* gully erosion, control of

infiltration
 effects on conservation design, 60, 68, 69
 effects on erodibility, 1
infiltration capacity
 effects on conservation design, 57, 61
 effects on erodibility, 21–2, 88
 effects on overland flow production, 7, 9
 relationship with plant cover, 25
 relationship with rainfall intensity, 16, 21,
  22
infiltration rate, 40, 56

Kirkby's mechanical sediment transport
 equation, 89, 95

land capability classification, 30–2, 40
land suitability classification, 30
land systems classification, 32, 36
land use, as factor affecting erosion, 1, 26, 72,
 79
 as factor in erosion hazard assessment, 37,
  40, 80, 83, 88
land use planning, as related to conservation
 strategies, 90, 95, 100–2
landslides, 11, 13, 17, 59, 66, 81
livestock, 5, 39, 59; *see also* rangeland

magnitude of erosion events, *see* frequency and
 magnitude of erosion
Manning equation, 8, 9
mass movements, 5, 11–13, 21, 37, 83; *see also*
 landslides
mechanical conservation measures, 57, 62–8,
 69, 88, 93
mechanization, 60
minimum tillage, 62
models, 41–2, 49–56, 80
mulch tillage, 62
mulching, 60–1

organic matter, 21, 22, 23, 57, 61, 86

overgrazing, 58
overland flow, 5, 13, 15, 20, 63
 as factor in erosion hazard assessment, 76,
  77, 79, 81, 83, 97, 99
 contribution to sediment removal from
  hillslopes, 6, 10
 effect of slope on, 25, 89
 energy of, 6
 frequency of, 17–18, 98
 hydraulics of, 7–9
 inclusion in erosion models, 54–5
 simulation in laboratory, 45

piping, 10, 11, 13; *see also* tunnel erosion
plant cover
 as factor affecting erosion, 1, 15, 20, 24
 as factor in erosion hazard assessment, 26,
  27, 39, 86
 effects on conservation design, 57–60, 61
 effects on erosion rates, 3, 25
 effects on overland flow production, 9
 inclusion in erosion models, 50, 56
 in Universal Soil-Loss Equation, 51

rainfall
 aggressiveness index, 27, 50, 73, 76, 77, 79,
  88, 99
 as factor affecting erosion, 1, 3
  in Universal Soil-Loss Equation, 51, 54
 drop-size distribution, 18–19, 45, 54
 intensity
  effects on conservation design, 62
  effects on overland flow production, 9
  relationship with erosivity indices, 19,
   20, 27, 28, 74–6
  relationship with infiltration capacity,
   16, 21, 22
  thresholds for erosion, 15–18
 simulation, 24, 45, 61
rainfall–run off relationships, 49, 89, 95
rainsplash erosion, 5, 76, 88, 89, 99
 as factor in erosion hazard assessment, 81,
  83
 contribution to sediment removal from
  hillslopes, 6, 10
 effect of slope on, 24–5
 inclusion in erosion models, 56
 processes of, 6–7
 see also detachment
range pitting, 63
rangeland, 39, 58
recreation, 30, 31, 59
relief
 as factor affecting erosion, 4, 103
 as factor in erosion hazard assessment, 37,
  39, 40, 84, 89
 inclusion in erosion models, 50
Reynolds number, 7, 9

rill erosion, 5, 44, 79
    as factor in erosion hazard assessment, 36,
        37, 39, 80, 81, 83, 89
    contribution to sediment removal from
        hillslopes, 6, 10
    effect of slope on, 25
    inclusion in erosion models, 54
    processes of, 10
    relationship with gullies, 11, 13
    relationship with overland flow, 9
    relationship with rainfall, 15, 17, 20, 76
ripping, 62, 68
rotation of crops, 57–8, 60
rotational grazing, 39, 58
roughness, 9, 25, 54, 57
run off
    as factor in erosion hazard assessment, 37,
        39, 72, 76, 83, 89
    effect of slope on, 24
    effects on conservation design, 57, 59–63,
        66–7, 78, 97, 101, 102
    field measurement of, 42, 44–5
    inclusion in erosion models, 56
    in Universal Soil-Loss Equation, 54
    relationship with soil loss, 28
    response to rainfall, 15–17, 21, 22, 88
    see also overland flow; rainfall–run off
        relationships

saltation, 13
sampling, 39, 44–5, 72–3, 86, 97
sand traps, 45
scale in erosion studies, 3–4, 26, 42, 69, 70, 103
sediment concentration, 3, 45, 49, 72
shear strength, 21, 86, 88
shelterbelts, 68
silt/clay ratio, 97–8
slaking, 7
slope
    as factor affecting erosion, 1, 15, 88
    as factor in erosion hazard assessment, 27,
        28, 37, 39, 80
    effects on conservation design, 60, 62, 63,
        66, 67, 69, 90–1, 100–1
    effects on soil loss, 24–5
    inclusion in erosion models, 49
    in Universal Soil-Loss Equation, 51–4
    maximum for acceptable soil loss, 89–90,
        95
slope maps, 83, 97
slope sampling, 44, 86
snowmelt, 17
soil
    as factor affecting erosion, 1

as factor in erosion hazard assessment, 37,
    81
conditioners, 62
conservation, 1, 4, 14, 26, 30, 51, 57–69,
    89, 90, 92, 103
management, 57, 61–2
see also aggregate stability; clay minerals;
    erodibility; organic content; silt/clay
    ratio; soil management
spatial variations in erosion, 3–4, 15, 40, 103
stabilization structures, 67–8
Stanford Sediment Model, 54–5
strip cropping, 60, 62, 63
strip-zone tillage, 62
subsoiling, 62
subsurface flow, 10, 11, 24, 61

temporal variations in erosion, 3, 15, 103
terrace channels, 63, 66, 67
terraces, 1, 40, 70, 83, 86, 95
    in conservation strategies, 69, 91, 96,
        100-2
    in Universal Soil-Loss Equation, 51–4
    types of, 63–6
tied ridging, 63
tile-outlet terraces, 67
tillage, 5, 21, 56, 61–2
timber exploitation, 58–9
trampling, 5, 24, 39
transport, of soil particles, 5, 6, 13, 21, 56, 57,
    103
transport capacity, 9, 14, 56
transport-limited condition, 5
tunnel erosion, 10, 11, 62, 68

Universal Soil-Loss Equation, 51–4

water erosion, 2, 6, 13, 23, 39, 54, 60
waterlogging, 60, 63
watershed conservation, 30, 90
weathering, 5, 11, 17, 56, 71, 97
weed control, 60, 61, 68
wind
    breaks, 1, 68
    erosion, 1, 23, 39, 54
        effects of plant cover on, 25
        effects on conservation design, 57, 60,
            61, 62, 68, 69
        field measurement of, 45
        processes of, 13–14
    erosivity, 20–1
    tunnels, 14, 23, 45, 48